AF557757

MICROORGANISMS

MICROORGANISMS

By

DR. S. SUNDARA RAJAN

Principal

Reva Institute for Science and Technology Studies

R.T. Nagar

Bangalore – 560 024

ANMOL PUBLICATIONS PVT. LTD.

NEW DELHI - 110 002 (INDIA)

ANMOL PUBLICATIONS PVT. LTD.
4374/4B, Ansari Road, Daryaganj
New Delhi - 110 002
Ph.: 3261597, 3278000
Visit us at: www.anmolbooks.com

Microorganisms

First Edition, 2003

ISBN 81-261-1391-X

PRINTED IN INDIA

Published by J.L. Kumar for Anmol Publications Pvt. Ltd., New Delhi - 110 002 and Printed at Mehra Offset Press, Delhi.

CONTENTS

Preface

The living world is most fascinatng to any obsever due to its wide diversity .While the manifest (macro) has attracted the attention of man even before the invention of the microscope, the unmanifest (micro) opend up its kaleidoscopic variety to the human eye when they (microbes) became obvious only through the microscope.

MICROORGANISMS provides a birds eye view of the fascinating microcosm. Starting with the histrory and invention of the microscope, the book in its 12 chapters provoides a general survey of all groups of microbes such as Viruses, Mycoplasma, Protozoa etc .

The book is meant to be a text book for students of life science, pharmacy, agriculture etc at the graduate and post graduate level .

I am thankful to Mr J.L Kumar, M.D of Anmol Publications, New Delhi for publishing this book.

Dr. S. SUNDARA RAJAN

Principal

Reva Institute For Science and Technology Studies

R.T. Nagar

Bangalore - 560 024

1

INTRODUCTION

Microbiology, the study of microrganisms, derives its name from three Greek words – *mikros* (small), *bios* (life) and *logos* (study). This means that microbiology deals with the study of microorganisms. What are microorganisms? Organisms which are so tiny and invisible to the naked eye constitute microorganisms (microbes). If any object is smaller than 0.1 mm, the human eye can not perceive it and even at a size of 1.0 mm very little details of an object can be seen with the naked eye. Hence, we can say that study of organisms with a size of 1 mm or less, comes under the perview of microbiology. Microorganisms can be looked into and studied only with help of a microscope.

Scientists opine that microbes originated on our planet about three or four billion years ago from complex organic materials present in ocean waters or possibly in vast cloud banks. As the first forms of life, they are regarded as ancestral to all forms of life on earth.

BEGINNING OF MICROBIOLOGY

Although microbes are most ancient and they had the planet all for themselves initially, even after the advent of man they have been influencing his life both for good and bad since time immemorial. In a lighter vein one can say that ever since the first toast was proposed and the first loaf of bread was baked, man has known the influence of microbes.

In spite of the above statement it seems incredible that the scientific study of microbes or microbiology is an infant science. In fact, but for one single instrument, viz., the microscope, the entire microbial world would have been unknown to us. Microbes were observed for the first time by Leeuwenhoek a little more than 300 years ago and even then their role in human life was never contemplated. They were just thought to be cute tiny animalcules and their study was a mere curiosity. There was, indeed, a lapse of more than 200 years since their discovery before they were studied seriously as notorious pathogens. Conclusive refutation of the theory of spontaneous generation by Louis Pasteur, and the germ theory of disease by Robert Koch may be mentioned as the two most significant discoveries that made many scientists truly open their eyes towards microorganisms and their role in human welfare. The discovery by Louis Pasteur that microbes are not the result but cause of fermentation opened up an entirely new vistas–microbes as agents of chemical change.

SCOPE OF MICROBIOLOGY

Since microbiology deals with a group of particular life forms it comes under the broad domain of biology which includes the study of all aspects of living beings including man.

Where can we fit in microbes in the hierarchy of living beings? Traditionally, living beings are divided into plants and animals. But members of microbes can be accommodated in both plants (fungi) and animals (protozoa) and some cannot be accommodated in either plants or animals as they share the characters of both. (For instance, *Euglena* was a disputed property till recently between botanists and zoologists.)

In one of the earlier attempts to resolve this problem, Haeckel (1866), a German Zoologist, suggested that there should be a third kingdom besides *Plantae* (plants) and *Animalia* (animals) to include all the microorganisms. He gave the name Protista to this kingdom to include all unicellular microorganisms that are neither plants nor animals.

Haeckel's classification raised some questions like how to distinguish a fungus from a bacterium or from an alga. The discovery of the prokaryotic and eukaryotic nature of the cells in late 1940s rendered the three kingdom classification unsatisfactory.

A recent and comprehensive classification proposed by R.H. Whittaker (1969) has five kingdoms of living beings :

Kingdom Monera
Kingdom Protista
Kingdom Fungi
Kingdom Animalia

Kingdom Plantae

Microorganisms include three (Monera, Protista and Fungi) of the five kingdoms mentioned above. At present it is agreed that within the preview of microbiology, five major groups of microorganisms–viruses, bacteria, fungi, algae and protozoa – are dealt with.

As will be evident from the above discussion , the scope of microbiology extends to both eukaryotic as well as prokaryotic microbes. While discussing the scope of microbiology it should be evident to us that it does not deal with merely the enumeration of structural diversity or classification, but extends to all aspects of microbial life. Microbiology is concerned with their form, structure, reproduction, physiology, metabolism, classification, and most important, their economic importance. In other words, what the microbes can do and should not be allowed to do (some times), as far as human beings are concerned, is one of the vital aspects of microbiology on which rests human destiny.

BRANCHES OF MICROBIOLOGY

It has been mentioned earlier that microbiology is not a mere study of the structural diversity and classification of microbes, but encompasses the whole gamut of microbial life. The knowledge of the various aspects of microbes has been accumulating since the last century and has become so vast that no microbiologist can claim familiarity with all aspects of the subject. The various aspects of microbiological study can be divided basically into the following branches.

1. Phycology

Deals with the study of autotrophic eukaryotic organisms. Members are generally called algae. Algae include both microscopic as well as macroscopic members. Only the microscopic algae are studied as a part of microbiology.

2. Mycology

The study of eukaryotic, achlorphylous organisms generally referred to as Fungi, is included in this branch. Some of the common fungi are yeasts, moulds, mushrooms, puffballs, etc. Fungi are not only harmful but beneficial also.

3. Virology

Viruses are neither eukaryotic not prokaryotic. In fact, they are on the border line between living and non-living. Viruses cause disease to plants and animals including human beings. The dreaded AIDS is also caused by a virus.

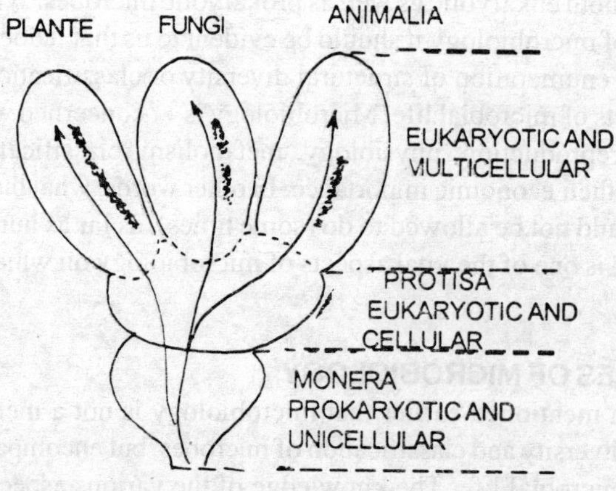

Fig. 1.1 Aim and Scope (Of Microbiology)
Five Kingdom classification of living beings

4. Protozoology

Study of Protozoans in all their aspects comes under the perview of protozoology. Protozoans are known to cause many diseases like malaria, amoebic dysentery, sleeping sickness, etc.

5. Bacteriology

This is the largest group among microbes not only in number but also in importance. Bacteria of both kinds–Eubacteria and Cyanobacteria (also known as blue green algae)–are studied here. Bacteria have a profound influence on various human endeavours including health, industry, agriculture, etc.

6. Medical microbiology

This branch deals with the pathogenic microbes–their life-cycle, physiology, genetics, reproduction etc.– many of the microbes also provide remedies for microbial diseases. All these aspects are studied in this branch. Some of the diseases like tuberculosis, leprosy, typhoid etc. are caused by microbes, and cure for them is provided by other microbes in the form of antibiotics.

7. Agricultural microbiology

In this branch, the role of microbes in agriculture is studied from the point of view of both harm and usefulness. Many microbes–fungi, bacteria and viruses–cause a number of plant diseases. From the point of view of benefit – N_2 fixing activity, use of microbes as biofertilizers and several other aspects are studied.

8. Industrial microbiology

The role of microbes in Industrial Production is studied here. Many microbes produce industrial alcohols, and acids as a part of their metabolism. The study of fermentation by microbes has contributed a lot to alcohol manufacturing. Breweries have greatly benefitted by understanding the role of specific microbes in fermentation.

9. Food and Dairy microbiology

Various aspects such as food processing, food preservation, canning, pasteurization of milk, study of food borne microbial diseases and their control is studied.

10. Aquatic microbiology

Microbiological examination of water, water purification, biological degradation of waste are studied in this branch.

11. Aero microbiology

Dispersal of disease causing microbes through air, microbial population in air, their quality and quantity in air comes under the perview of this branch.

12. Environmental microbiology

This is one of the most important branches of microbiology. The role of microbes in maintaining the quality of the environment is studied here. Microbial influence in degradation and decay of natural waste, their role in biogeochemical cycles are all studied. Some of the recent researches have shown that certain bacteria can help in cleaning the oil spill, and this gives added significance to the study of environmental microbiology.

13. Geochemical microbiology

Role of microbes is coal, gas and mineral formation, prospecting for coal, oil and gas and recovery of minerals from low grade ores using microbes, is included here.

14. Biotechnology

This is the most significant branch which may even change the course of life as we know today. Microbes are used as gene carriers to deliver specific genes to function in a different environment. New, genetically engineered microbes can produce drugs (human insulin), or in agriculture-N_2 fixing ability may be transferred to all the plants. The potentialities of biotechnology are immense.

15. Immunology

Studied in this branch are the immune responses in organisms. How toxins are produced? How the antigens influence the formation of antibodies? How protective vaccination helps in combating the diseases? How immune system collapses (as in AIDS), are some of the questions for which immunology as a branch of microbiology is trying to find out answers.

16. Exo microbiology

This is a branch still in its infancy. Study of life in outer space comes under its perview.

IMPORTANCE OF MICROBIOLOGY (Role of Microbiology in Human Welfare)

Why should we study microbiology? Has there been any change in the quality of our life due to knowledge that we have acquired from the study of microbes? In other words, what is the importance of microbiology in our life? We will try to find answers to some of these in the discussion below.

The discussion on the role of microbes in human welfare may be divided under two headings – good and bad. Microbes, as we know, are capable of both good and bad as far as human life is concerned. We will now list both the harms and benefits by microbes and then draw a conclusion as to how microbiology has helped us to control or kill the bad microbes and make maximum use of good microbes.

Harmful effects

Many of the microbes may be regarded as man's worst enemies. They cause a wide array of diseases that not only threaten the existence of plants and animals on which he depends, but also his own existence directly. Viral, bacterial and fungal diseases have taken a heavy toll of human life, destruction of domesticated animals and total loss of agricultural crops. Before the discovery of microbes, and before they were connected to diseases (germ theory of disease), the reasons for these diseases were not known and man continued to suffer. But, the development of microbiology gave man an insight into the life of microbes. Study of the structure and life history of microbes and their physiology, metabolism, genetics etc. will help us to control them so that diseases can be prevented and if it occurs can be cured.

Beneficial activities

The benefit that man can derive from the activities of microbes are immense. Study of these in microbiology help us to improve (both in quality and quantity) these activities so that man can be benefitted. Medical microbiology, Agricultural microbiology, Environmental microbiology, Industrial microbiology, biotechnology–are all branches of microbiology that tell us as to how in various human endeavours, microbiology is giving a helping hand to improve the quality of life.

Biotechnology is one field which has the potential to change the course of life itself. Employed properly, it can bring about vast improvement in human life and the environment. But, if used improperly (just as atomic energy is used for making bombs), it can bring untold misery to human life.

Creation of new genetically-engineered microbes using the technique of Recombinant DNA, may help us to produce new drugs, disease-resistant animals and crop plants etc. Some microbes may also help us to check pollution, as for example, bacteria that can feed on oil spill.
As research tools, to enquire into the fundamental processes of life, microbes have rendered great service. They can be cultured, life histories can be studied in a short time, and the results obtained can be interpreted for higher forms of life (including human beings) also, as the genetic material is same.

From the discussion above, it must be abundantly clear that microbiology has become increasingly important to human society. It (microbiology) has emerged as one of the most important branches of life sciences. As microbes practically affect all activities of our life, like food, clothing, shelter, health, hygiene etc. Microbiology also has made vast progressive strides in all these fields in little less than a century to improve the quality of our life. Infectious diseases have almost been conquered by new drugs, quality of agricultural crops improved by using techniques of genetic engineering, new varieties of wines, liquours have been produced perhaps to raise the spirit of man – all these are possible only because of microbiology. All these will make us wonder, how our life would have been without the knowledge of microbiology?

2

HISTORY OF MICROBIOLOGY

Life originated on this planet some three or four billion years ago. Since then it is manifesting itself in various forms through the process of evolution. The first few steps that the nascent life must have passed through initialy in its never ending quest for diversification must have been in the form of microbes. Microbes must be as old as life itself and their influence on man is older than the recorded history of science. The study of microbes is known as microbiology (*Micros* = small; *logos* = study).

Even though man has been experiencing the influence of microbes since time immemorial, the scientific study of microbes is comparatively recent. The science of microbiology was unknown before the discovery of that most important tool of microbiologists – the microscope.

From the first century BC it was believed that diseases are caused by invisible beings. During the 13th Century, Roger Bacon (1220–1292), suggested that diseases are caused by invisible beings. The opinion was supported by Girolamo Fracestoro Verona (1483–1553) and Anton von Plenciz in 1762. Earlier than Anton von Plenciz, a monk – Anthanasius Kirchers – related the decay of bodies, meat, milk etc., in invisible worms.

The most important discovery that may be called as the founding step of the science of microbiology was that of Antoni van Leeuwenhoek, a Dutch scientist who lived in Holland from (1632–1723). With his crude optical device, which is the forerunner of the modern day microscope, he observed tiny microbes in a drop of pond water which he called animalcules. These are now regarded as bacteria. Since then started the glorious era of microbiology.

In the history of microbiology three specific phases may be recognised . These are :

(i) Discovery phase
(ii) Transition phase, and
(iii) The golden age

During the discovery phase, the microbes were discovered and attention was devoted to mainly observe and describe the microbes. The names of Marcello Malphigi and Antony van Leeuwenhock are important during this period.

The transition period in the development of microbiology coincided with a great controversy regarding the origin of microbes. The controversy arose as to the opinion of some, that life arises spontaneous! From a non-living source (abigonetic origin of life), as one can see the sudden appearance of tiny maggots, worms and flies as if from nowhere on any decaying organic matter. The experiments of Fransisco Rhedi, John Needham, Spallanzani and Nicolas Apperte are worth mentioning.

The third phase in the history of microbiology, which may be referred to as the golden age, began in 1860 and continued up to 1910 when many important contributions were made. The most significant contribution during this period came from Louis Pasteur and Robert Koch.

The following table summarizes the important discoveries in the field of microbiology.

	Name of the Scientist	Year	Discovery
1.	Anthanasius Kircher	(1601-1680)	Invisible worms cause diseases.
2.	Girolamo Francastoro	(1483-1553)	Invisible beings cause diseases
3.	Francesco Rheedi	(1626-1697)	Disproved the theory of spontaneous generation
4.	Antony van Leeuwenhoek	(1632-1723)	Pioneer in the discovery of microbes
5.	John Needham	(1713-1781)	Conducted experiments which seemed to prove the theory of spontaneous generation.
6.	Lazaro Spallanzani	(1729-1799)	Experimentally disproved the

		theory of spontaneous generation
7. Edward Jenner	(1749-1823)	Discovered vaccine against small pox
8. Justus von Leibeg	(1803-1873)	Discovered chemical basis of fermentation.
9. Jacob Henle	(1809-1885)	Established the fundamental principles of the germ theory of diseases.
10. Louis Pasteur	(1822-1895)	Discovered fermentation by living cells; conclusively disproved the theory of spontaneous generation; developed rabies vaccine.
11. Joseph Lister	(1827-1912)	Introduced antiseptic methods in surgery
12. T.J. Burill	(1839-1916)	Discovered bacterial plant diseases
13. John Tyndall	(1839-1916)	Discovered fractional killing of bacterial spores.
14. Robert Koch	(1843-1910)	Proposed Koch's postulates for germ based diseases; identified tuberculosis causing bacilli.
15. Paul Ehrlich	(1854-1915)	Developed the principles of Chemotherapy to cure diseases.
16. Elie Metchnikoff	(1845-1916)	Discovered the phenomenon of phagocytosis.
17. Hans Christian Gram	(1853-1933)	Discovered differential staining of bacteria using gentian violet dye.
18. N. Winogradsky	(1856-1934)	Discovered N_2 fixing bacteria in the soil.
19. William Welch	(1850-1934)	Discovered the relation of anaerobic bacteria to gangrene.
20. Walter Reed	(1851-1902)	Discovered the transmission of yellow fever by mosquitoes.
21. Theobald Smith	(1909)	Discovered rickettisiae.
22. Theobald Smith	(1859-1934)	Discovered that Texas fever is transmitted by ticks.
23. Ronald Ross	(1897)	Discovered the sexual cycle of malaria parasitie in mosquito.

Name of the Scientist	Year	Discovery
24. Jules Bordet	(1906)	Isolated the whooping cough bacillus.
25. David Bruce		Discovered that sleeping sickness disease causing microbe is transmitted by tse tse fly.
26. George Gafky		Isolated typhoid bacillus.
27. Shiba saburo Kitasato		Isolated teanus bacillus.
28. Emil von Behring		Development of anti diptheria toxin.

We will now study in some detail the contributions of the following scientists.

ANTONY VAN LEEUWENHOEK (1632-1723)

Leeuwenhoek, who lived from (1632-1723), is credited as the discoverer of the microbial world. He was a draper and haberdasher, and owned a small dry goods shop in Delft, Holland, where he sold clothes to the wealthy ladies of the city. He was also a qualified surveyor and official wine taster of the town. As a hobby, he used to grind glass and make lenses. He often used a magnifying glass to study the weaves of the various types of cloth. But since the lenses were not so good, Leeuwenhoek set himself up to grind more perfect lenses. He fixed his lenses placing them between two silver or brass plates rivetted together. The opening between the plates was less than 1/16 of an inch. The specimen to be studied was mounted on metal point in front of the screw. During his life time Leeuwenhoek constructed more than 200 such microscopes. He was so successful in grinding lenses that he gave up textile business and took to full time lens making. The 'microscopes' of Leeuwenhoek could magnify objects about 200–300 times.

With his microscope, Leeuwenhoek observed a variety of things mainly out of curiosity. Hair fibres, plant structures, crystals, insect, insect's eye, a variety of fluids such as pond water, blood etc, scrapings from his own teeth and transparent tail fins of fish were observed by him.

For some 50 years Leeuwenhoek kept meticulous records and made accurate drawings of his observations. He examined blood and discovered some tiny microbes which are nothing but erythrocytes. He examined yeasts and found that they were made up of tiny round particles.

Leeuwenhoek's greatest claim to frame in the field of microbiology however, was his discovery and description of microbes, which he called animalcules. He found them in a variety of sources like rain water, pond water, pepper grain infusion, scraping from his own teeth etc. He was astonished to see them alive, moving to and fro in the field of his microscope. All the main types of unicellular microorganisms – protozoa, algae, yeasts and bacteria were first described by him as early as in 1676.

Leeuwenhoek had been corresponding with the British Royal Society and used to send communications regarding his microscopic observations. In one of the first letters dated Sept. 7th 1674, he described "very little animalcules", which we now identify as protozoa. His eighth letter which he wrote on October 9, 1676, contained detailed description of the microbes. The descriptions were accoimpanied by accurate sketches. In his own words–

Fig. 2.1 History of Microbiology
Antony Van Leeuwenhoek, the discoverer of microbes

"In the year 1675, I discovered living creatures in rain water which had stood but a few days in a new earthen pot, glazed blue within. This invited me to view this water with great attention, especially those little animals appearing to me ten thousand times less than those …. which may be perceived in the water with naked eye".

He was accurate in his descriptions and we know for sure that what he observed must have been a variety of fungi, algae, protozoa, bacteria etc. On June 16, 1675, he reported that while examining well water into which he had put a pepper grain ….. "I discovered in a tiny drop of water, incredibly many little animalcules and these of diverse sorts and sizes. They moved with bendings, as an eel always swims with its head in front and never, tail first, yet these animalcules swam as well backwards as forwards, though their motion was very slow".

Obviously, what Leeuwenhoek saw were the mobile bacteria. In 1683, he described and sketched different forms of animalcules-rods, spheres and spiral shapes which are nothing but the morphological forms of bacteria. This was the first recorded observation of bacteria and even today his morphological classification of animalcules holds good for bacteria.

Leeuwenhoek, however, did not go beyond describing the microbes. He never tried to associate them with their surroundings as causative agents. Also, nor did he reveal the secret of his technique of grinding lenses. During the time of Leeuwerthoek and even later, microscopic observation was regarded as an idle hobby with no practical relevance.

EDWARD JENNER (1749-1823)

Edward Jenner is credited with the discovery of a safe and efficient method of immunization against small pox. Jenner, an English country physician, developed the method of vaccination against small pox virus. Vaccination deals with the introduction into human skin of the cow pox causing virus (Vaccinia), cow pox is a cattle disease that commonly appeared (in Jenner's time) in the form of pustules on the teats. Similar pustules or lesions sometimes developed on the hands of milkmen who handled infected cows for milking. These cow pox lesions were very much similar to those of small pox, but they were localized and caused mild symptoms of the disease without any complications. Additionally, cow pox was not communicable from one person to another. Another common observation made by the local people was that a person who once contracted cow pox will be totally immune from small pox and some how the cow pox lesion provided protection against small pox.

As a medical student, Jenner had keenly observed these facts and beliefs, and as a physician he started serious investigations. After several years of study he was convinced that cow pox lesions do offer protection against small pox. In 1796, Jenner collected the fluid from a cow pox pustule from the hand of one Sarah Nelms, a dairy maid, and transferred it on to the skin of a healthy boy, James Philip. The boy developed a typical lesion of what we now get after a vaccination. Several weeks later, Jenner inoculated this boy from the fluid taken from the pustule of a small pox affected individual, but the disease (small pox) did not appear. Thus the boy had immunity from small pox. Subsequent studies and data collection by Jenner from similar experiments clearly proved that inoculation with cow pox fluid would indeed offer immunity from small pox. Jenner published his findings in June 1798, in the now famous pamphlet. An Inquiry into the cause and effect of variole vaccinae. This may be truly regarded as one of the greatest achievements ever made by man in his quest for conquering of disease and is also heralded as a significant milestone in the history of microbiology. Though vaccination was met with much opposition first, it has stood the test of time and proved its usefulness (Note: Small pox virus, however, has been completely eradicated from the earth and immunization against it is no longer necessary).

LAZARO SPALLANZANI (1729 – 1799)

The name of Spallanzani is associated with those microbiologists who disproved the spontaneous generation theory (Albiogenetic origin of life). In order to understand this, we must understand in some detail the theory of spontaneous generation.

The theory of spontaneous generation

After the path breaking discovery of microbes by Leeuwenhoek, scientists began to ponder over the origin of the microbes. From the beginning there were two schools of thought. Some believed that the animalcules (microbes) were produced spontaneously (without any external influence) from non-living materials, whereas, others thought that the formation of animalcules on the decaying matter is nothing but their growth on a suitable medium and that the animalcules were always present in air (invisible to the eye). The theory, that new forms of life can arise spontaneously from a non-living source, is called the theory of spontaneous generation or the abiogenetic origin of life. On the other hand, the theory that states that life can come only from pre-existing life is called the biogenetic origin of life.

The abiogenetic theory was experimentally disproved by Fransisco Rhedi, who lived a century earlier than Spallanzani. John Needham, an English priest performed experiments with mutton gravy that seemed to support the theory of spontaneous generation.

Spallanzani, an Italian priest and a naturalist took upon himself the task of supporting Rhedi's view and refuting the experiemental results of Needham. He conducted a long series of experiments. He set up four groups of flasks –

(i) Flasks made air tight by melting the glass neck of the flask.
(ii) Flasks closed with cotton plugs.
(iii) Flasks closed with wooden corks (similar to the ones used in Needham's experiment).
(iv) Flasks open to the air.

All the flasks were filled with seeds and vegetable matter and then heated for one hour at the beginning of the experiment. Spallanzani set them aside for 25 days and than examined the contents of the flasks microscopically. The open flasks were full of teeming life. Corked flasks also had thousands of animalcules in the same way as in Needham's experiment, but the flasks stoppered with cotton and hermetically sealed ones had very few animalcules.

Spallanzani concluded that the number of animalcules in a flask increased due to the exposure to air. The microbes in Needham's flask might have come through the air as the cork stopper may not have completely prevented the entry of air. But, what about the animalcules (small in numbers) that appeared even in the hermetically sealed flasks also? The flasks, however, did not develop any microbes initially, but later, as discovered by Spallanzani, a tiny crack had developed which allowed the entry of air and the development of microbes. Spallanzani's final conclusion was that to keep the infusion (vegetable matter) completely barren it should be boiled and hermetically sealed. This explanation did not satisfy the proponents of spontaneous generation and some quickly discredited Spallanzani's experiment of sealed flasks as there was no air. There could not be any immediate answer to this. Faculty experiments continued to be conducted and results brought forward to prove the theory of spontaneous generation. The controversy of abiogenesis lasted well over into the middle of the nineteenth century.

LOUIS PASTEUR (1822 – 1895)

Louis Pasteur was born in the village of Dole, in France, on December 27, 1822. He had a humble origin. His father was a tanner. Pasteur was originally trained as a chemist but later switched over to medicinal research. He obtained a doctorate in Chemistry from the Normal School in Paris. He taught at Strassburg and Lille before returning to Paris where he became a professor at the Eloledes Beaux Arts. In recognition of his pathbreaking discoveries, a special Institute – The Pasteur Institute was built for him by public support. Hailed the world over for his epoch making discoveries, Pasteur died in Paris in September 28, 1895. His tomb is in the basement of the Institute in Paris.

Pasteur's contribution to the development of microbiology are enormous, and he may be rightly called the father of microbiology. His main contributions are –

1. Tartaric acid, an organic compound, is formed by two types of crystals which could be separated microscopically.

2. ***Discovery of microbial fermentation*** : Theodor Schwann (1837) had demonstrated that yeasts were responsible for fermentation (the process in which carbohydrates turn into alcohol). This was against the opinion of Justin Leibeg, a German chemist, who thought of fermentation as a chemical process.

Fig 2.2 History of Microbiology
louis Pasteur

Pasteur was attracted to this controversy and wanted to investigate as to why wines turn sour. Pasteur observed under the microscope yeast cells and some bacteria. He believed that yeasts cause fermentation. In a series of experiments he showed that grape juice yeast mixture turns into alcohol. He killed the yeast cells by applying heat and showed that alcohol is not formed in grape juice. When yeasts are added again, alcohol formation reappears.He also showed that the bacteria are responsible for turning the wine wour.

The microbial fermentation was indeed a fundamental discovery, because, till then microbes were simply the objects of curiosity and nobody linked them with any chemical activity of significance. Pasteur's work also showed that microbial chemical activity may be the cause of many diseases unexplained so far.

3. Disproving the theory of spontaneous generation

Pasteur was convinced that the spontaneous generation of living beings is wrong. He believed that microbes are present in soil, air and water and when they get a suitable environment they grow. In other words, their spread can be controlled by adopting suitable measures.

In order to answer the criticisms against Spallanzani's experiment that in the absence of air, microbes are not produced. He used gun cotton as a stopper on the flasks containing sterile (heated) broth. Microbes did not appear until the cotton plug was removed.

In order to conclusively disprove the spontaneous generation theory, Pasteur invented the swan necked flasks. This flask allows free entrance to air while all dust and contaminents settle in the bend of the neck. Heated broth in these types of flasks remain, microoe free. If the broth was tipped into the neck and flown back, microbes appear again. This expirement finally dispelled the theory of spontaneous generation.

4. Discovery of aerobic and anaerobic microbes

Microbes that require oxygen to grow are called aerobes. Since animals require oxygen for growth, scientists in the middle of the 19th century, assumed that microbes also require oxygen. However, Pasteur's study of the process of butyric acid fermentation, revealed that microbes can survive in the absence of oxygen. Microscopic observation of the fermented product showed actively moving microbes. The mobility of the microbes however stopped when they were exposed to air. Pasteur also observed that if the broth is aerated, fermentation would stop. He coined the term *anaerobic* to refer to organisms that do not require oxygen.

5. Work silkworm disease

In 1865, Pasteur was asked to find the cause of *pebrine,* a disease affecting silkworms. Pasteur demonstrated that the disease was caused by a protozoan microbe and by choosing microbe free egg for breeding, the disease could be controlled. This was a significant step towards the establishment of the germ theory of disease and rabies.

6. Work on anthrax and rabies disease

Pasteur demonstrated that the anthrax disease in sheep are were caused by microbial rods (bacteria), and cultured them in sterilized yeast water and showed that the inoculation of culture would cause the disease in healthy animals. He also discovered the principle of protective inoculation or vaccination against the disease. He showed that inoculation of weekened microbes to animals would provide them with immunity and they would not contract the disease–thus was discovered the anthrax vaccine.

Pasteur successfully applied the principle of protective vaccination to another dangerous human disease – rabies (hydrophobia), caused due to mad dog bite. Pasteur, however, could not observe the causal agent (virus) under his microscope. But, he could develop a vaccine of weakened organisms (viruses) that provided immunity against this near fatal disease.

7. Pasteurization

The process of partial killing of microbes called pasteurization that is now widely employed in dairies was originally devised by Pasteur to prevent spoilage of wine and beer.

JOSEPH LISTER (1827 – 1912)

Joseph Lister was born in England on April 5, 1827 to a well to do quaker parents. His father also dabbled in science and had made several contributions to the improvement of the microscope. Lister obtained his degree in medicine in 1852 and then went on to become professor of surgery at Glassgow and Edinburgh and later at the King's College, London.

Lister's contribution to surgery are many, but his major contribution has been in preventing wound infection after surgery by following antiseptic methods. Lister, as a surgeon, was moved by the post-operative complications in patients due to wound infection and set upon to study the reasons for it. He published his research papers on this aspect in 1867. He observed a similarity between the wound inflammation and putrefaction and fermentation which Pasteur had already shown to be due to microorganisms. This lead him to the conclusion that microbial growth may be the cause of wound infection. He was sure that if he would protect the wound with dressing that killed the microbes, inflammation would never occur. This was the basis of antiseptic technique.

Lister chose carbolic acid (a well known preservative), as the microbicidial agent. He applied carbolic acid soaked dressing to cover the wound of compound fractures which usually caused (often fatal) inflammation. The results of the application of carbolic acid dressing were dramatic; wounds healed in double-quick time. Lister applied this technique to many fastering wounds and obtained similar results. Finally, Lister developed a routine for the use of carbolic acid during operations which came to be known as, *Lister's antiseptic system.*

Lister also suffered many criticisms, but he never got disheartened, and by the year 1875 he was honoured everywhere. The result of antiseptic methods of surgery was beyond words. It saved countless lives from the jaws of death due to wound infections.

(In modern days, *aseptic* and *not antiseptic* method is used *i.e.* the instrument and the environment is made germ free even before surgery and not just after surgery).

ROBERT KOCH (1843 – 1910)

Robert Koch was born on December 11, 1843 in Germany. He took his medical degree in 1866 and began practising but switched over to microscopic studies which engaged him full time. On April 22, 1876, Koch wrote to Ferdinand Cohn, a professor of botany about his study on anthrax bacillus. Cohn invited him to present his work at the university. Koch's paper on Anthrax disease was published in 1876 and he became well known in the medical world. Koch, later became professor of hygiene and Director at the Institute of Infective Disease at Berlin. He died in May 27, 1910. His main contributions are –

Germ theory of disease- work an anthrax bacillus

Koch observed that animals which died of anthrax had thread like organisms in their blood. Anthrax disease was a serious disease that brought untold misery to farmers of Germany. Even though many scientists earlier had reported the occurrence of bacilli in the blood of infected animals nobody tried to link them to the disease.

Koch, in order to find out the cause of the disease, injected the blood of dead sheep into the laboratory mice (He did not want to kill sheep, hence he did not use them as experimental animals). When a mouse died, he took its blood and injected it into a healthy mouse. Koch repeated this experiment 20 times, and each time the healthy mice died. Examination of the dead mice revealed swollen spleens and the presence of bacilli in their blood.

Koch tried to grow these bacilli in various broth cultures and finally succeeded in culturing and isolating them in the aqueous humour (liquid from eye) from the eyes of a cow. He placed a drop of the liquid on a depression slide and inoculated with a piece of infected spleen. Very soon he observed the multiplication of bacteria and also spore formation. He then transferred these spores to fresh vitreous humour and saw them germinate into vegetative cells. In order to conclusively prove the role of bacteria in the disease he injected the spores into the mice and saw them (mice) become victims of the disease. Examination of the dead mices showed the bacilli in their blood. This lead Koch to propose his 'germ theory of–disease' which states that microbes cause diseases.

Fig 2.3 History of Microbiology
Robert Koch

Koch's postulates : To confirm that a microorganism is the cause of a disease, Koch summarized an experimented method in a set of rules which are popularly known as Koch's postulates. According to these postulates, a microbe may be accepted as a causative agent of an infectious disease only if it satisfies the following conditions.

1. The microbe must be present in every single instance of the disease and under conditions which explain pathological changes and clinical features.
2. The microbe must be isolated from the infected individual and should be grown in pure culture.
3. The microbe obtained from the culture should produce the same disease when injected into a healthy animal.

4. The same microbes must be obtained from the body of the newly infected (after inoculation) animal.

Other contributions of Koch

a. *Discovery of tuberculosis bacilli* : Koch showed that all forms of tuberculosis were caused by the same bacillus – *Mycobacterium tuberculosis.* He showed that animals would develop the disease when inoculated with the pure cultures of the bacilli (satisfying all the postulates).

b. Koch introduced the method of making smears of bacteria on glass slides, and staining them with aniline dyes.

c. *Plate method for isolating pure cultures* : Koch, accidentally observed that a slice of potato had colonies of bacteria distinct from one another. Each colony had thousands of bacilli. It then occurred to him that bacteria could grow on a solid surface and it would be easy to isolate and grow the colonies separately on a solid medium. When separate colonies grow on a solid medium it will be easy to pick up only one type of a microbe, culture it and introduce it into the experimental animal. This was a great breakthrough in the study of infectious microbes, as it is difficult to separate microbes in a liquid medium. In his early experiments, Koch used gelatin as a solidifying agent to provide a firm surface. Subsequently, Dr. Walter Hesse and his wife Fanny used agar instead of gelatin.

This method of obtaining pure culture is used even today. The plate cultures are made on a specially devised glass plate called petri dish. This was first introduced by Petri (1887), one of Koch's assistants.

ALEXANDER FLEMMING

One early morning, on September 1928, a grey haired English scientist stepped into his laboratory at St. Mary's Hospital in London. Never did he imagine even in his wildest dreams that destiny walked along with him into the laboratory that pleasant morning. He was Alexander Flemming – the discoveror of the first of wonder drugs - antibiotics.

Flemming, a trained bacteriologist and a student of Almroth Wright (the investigator who described opsonins), had been cultivating various bacteria in petri dishes laid out on laboratory tables. Like a curious gardener impatiently waiting for the new seedlings to sprout their heads, Flemming strolled among the tables and examined his microbe nursery. One of the Petri plates immediately attracted the attention of Flemming, for the bacterial culture there had been contaminated. Instead of throwing off the petri plate (as any ordinary person would have done), Flemming examined it thoroughly.

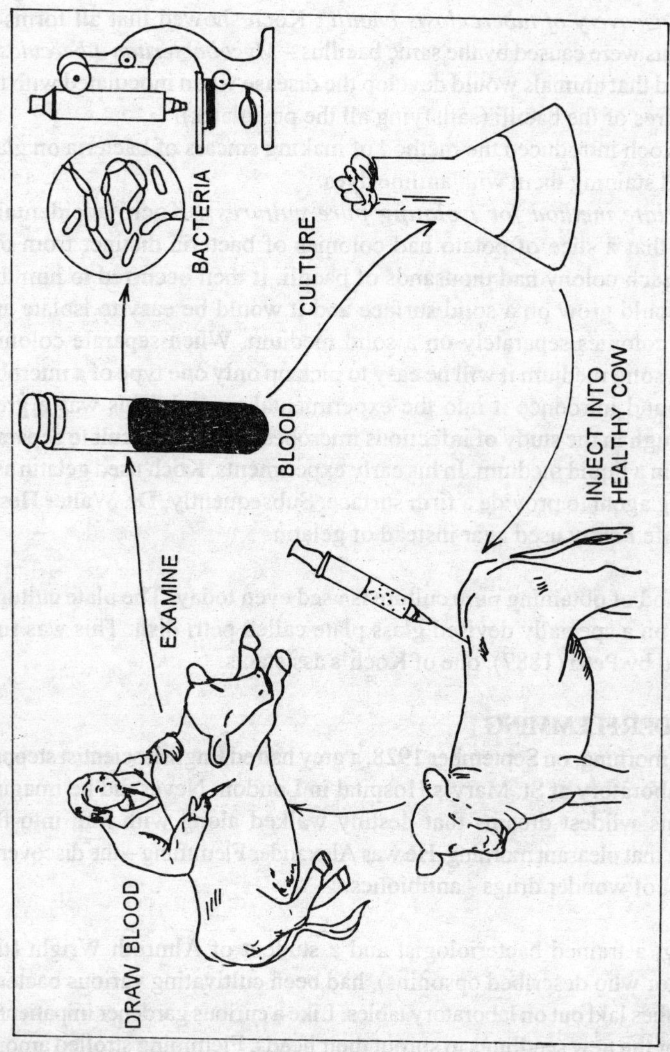

Fig. 2.4 History of Microbiology
Koch's postulates

Something strange had happened to the petri plate containing a culture of a golden type of *Staphylococus* bacilli. In the middle of the culture, a cluster of blue green mould had appeared overnight, perhaps a spore carried on by some vagrant breeze (or is it the hand of destiny), an invisible speck buoyed and wafted by unseen air currents had landed itself on that petri dish through a tiny opening, in what can be called a golden accident of history. Flemming was not interested in the appearance of the intruder (the mould), but what it had done to his bacterial culture. The bacterial culture had an odd appearance surrounding the mould. A clear transparent ring surrounded the mould, like a halo, while other parts of the culture had the glistening murky or cloudy appearaence that confirmed the presence of millions of bacteria in a splendid state of health. But, within the crystal clear ring all the germ life (bacilli) had been wiped clear as if by a wild fire epidemic. To the trained mind of Flemming, it appeared that something that is excreted by the intruding mould proved deadly to the germs and had annihilated all of them.

The intruding mould was then identified as a common blue green mould *Penicillium notatum* that contaminates clothes, shoes, fruits, bread and cheese. Looking like a broom or brush (Gr. *Penicillus* = brush) under the microscope, this ordinary member of the mould family was thought as a nuisance than of any use.

Curious at first and deeply interested later, Flemming cultivated his little broom mould and obtained from it a little broth (exudate). In order to test his hypothesis that the mould secretions can kill *Staphylococci,* he injected this mysterious juice (broth) into the laboratory mice infected with *Staphylococci, Sterptococci* and *Pneumococci.* Sure enough, the mysterious something in the juice killed all the germs in the body of the mice as it had done on the Petri dish. Flemming named this 'something' *Penicillin* even though he did not isolate it in pure form. In June 1929, Flemming published a paper in the British Journal of Experimental Pathology. In this paper, Flemming wrote "Penicillin may be an efficient antiseptic for injection into the areas infected with penicillin sensitive microbes"

This exciting news, however, did not create the sensation that it should have. It lay forgotten for many years until Howard Florey in association with Ernst Chain, a refugee German biochemist, was stimulated by a rereading of Flemmings, paper and was successful in obtaining a crude of penicillin – less than a level of tea spoonfull in quantity.

The first injection of Penicillin was given to a London boy who had nicked himself while shaving and had acquired a staphylococcal infection. His condition dramatically improved, but the world's supply of Penicillin (less than a spoonful) got exhausted and the patient could not be saved.

Since, conditions in Britain at that time were not suitable for any funding for research, Americans, attracted by the reports on penicillin invited Florey and Chain to work on Penicillin. Working in the USA, they were able to produce large quantities of Penicillin for trial and subsequent use. In the year 1945, for the discovery of Penicillin, Flemming, Florey and Chain shared the Nobel Prize in physiology and medicine (Flemming however was not involved in the work done in USA on Penicillin).

IWANOWSKY

The name of Dimitri Iwanowsky occupies a pivotal position in the history of virology. The casual agent of the Tobacco mosaic disease was unknown till then. Earier to Iwanowsky, Adolf Meyer, a Dutch agricultural chemist had shown that the sap from the infected leaves could cause disease to the healthy plants. He also demonstrated that the sap even after being filtered through filter paper retained the infectivity. He concluded that filterable bacteria cause the disease.

Iwanowsky (1892), was attracted towards this problem. He obtained the sap from the infected leaves and passed it through a bacterial filter (Chamberland candle filter), which filters out all bacteria. The filtered sap still retained the infectivity even though it had no bacteria. However, the culture of the filtered sap did not yield in any living organisms. Hence he concluded that microbes which cannot be cultured on a medium, but which are smaller than bacteria cause the disease. He, however, could not isolate Viruses which are the real causes of the disease.

3

MICROSCOPY

The study of cell biology involves the use of a number of instruments. The rapid strides made by cell biology would not have been possible, but for the pivotal role played by these instruments in untraveling the mysterious world of microcosm. For instance, if there were to be no microscope, the whole of the microbial world would be non available to us and we would not be in a position to explain the various diseases that confront us; for we would not know what microbes are. In this chapter we wiil try to understand the various instruments and their working mechanism. Microbiological instruments may be basically categorized into two – those that are used for visual observation and those that are used for analytical purposes. Visual observation of microbes is made possible by the use of microscopes while a wide array of instruments such as spectrophotometers, centrifuges, nephelometers, autoclaves inoculation chambers etc., are used for various analytical purposes.

In this chapter we shall study the various optical instruments (microscopes) used in cell biological study.

Microscopy

The most important tool of a biologist is the microscope. The invention of light microscope, in the sixteenth century, is perhaps the single most important contribution ever made to the advancement of biology.

The present day microscope used in biological laboratories has been developed over the course of several centuries. Ancient Romans, Greeks and Indians had known lenses of glass and quartz. In the 14th century itself, spectacles and single lenses were used as magnifiers. Zacharias Jensen, a Dutch astronomer, added one more lens and constructed a true telescope. Campini, an Italian scientist found out that lens can be ground to the desired curvature. Since the curvature of the lens increases its magnifying power, microscope thereafter, became a very powerful tool in the hands of biologists.

Antony Van Leeuwenhoek (1632 - 1723), some times wrongly credited with the invention of the microscope, used his single lens microscope to observe several minute objects like insects, mold, pond water, saliva etc. It was Robert Hooke (1665), an Englishman, who developed an instrument that could truly be referred to as the forerunner of the modern day microscope.

The microscope opened up to biologists (hitherto unknown to them) an exciting microcosm – the universe of microorganisms. Bacteria, cells, detailed structure of tissues, all became visible under the microscope. New fields of biology like cytology, embryology owe their existence to the single most powerful tool of biology – the microscope.

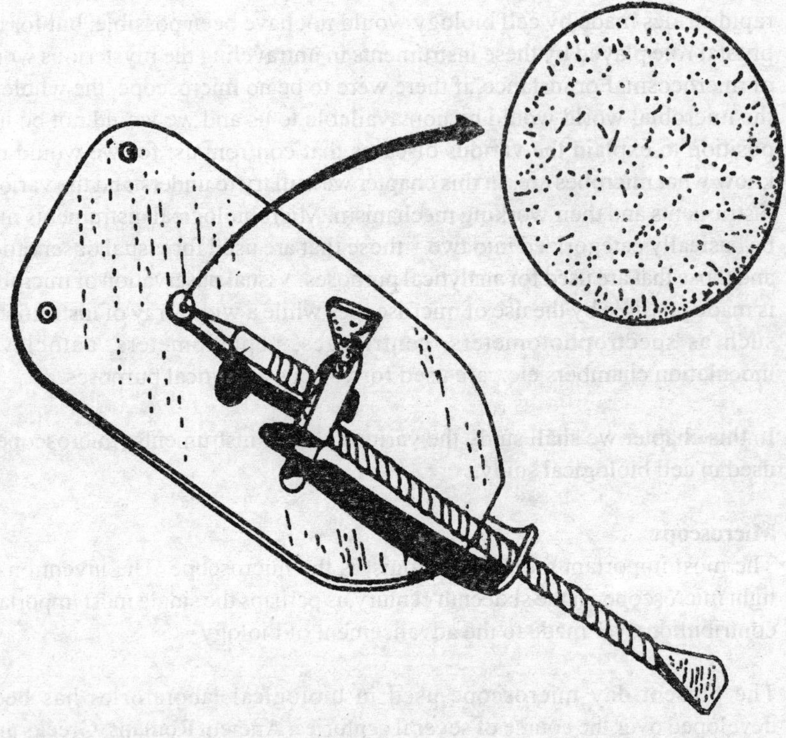

Fig. 3.1 Microscopy

Van Leeunhoek's Microscope. A simple lens of high magnification, this was the forerunner of the modern microscope.

The working of a microscope

The main function of the microscope is to produce a large and clear image of the object. Clarity is very often referred to as the resolving power of the microscope.

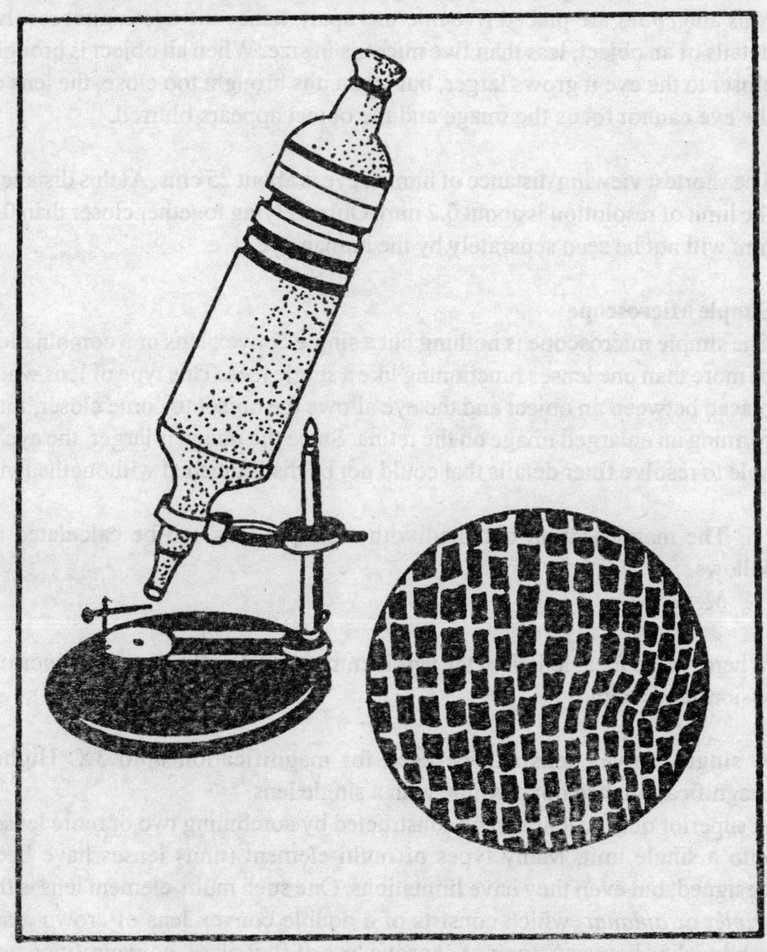

Fig. 3.2 Microscopy
Hooke's Microscope and cork cells

In order to understand the functioning of the microscope, it will be necessary to understand as to how our own eye resolves and produces an image of the object.

The lens of the eye focuses the light reflected by an object and projects an image on the retina. The retina is a granular matrix consisting of light sensitive cells called rods and cones. These cells convert the light impulses falling on them into nerve impulses producing a sensation of vision in the brain. The *rods* and *cones* are placed five microns apart, hence the eye cannot resolve details of an object, less than five microns in size. When an object is brought closer to the eye it grows larger, but when it is brought too close, the lens of the eye cannot focus the image and the object appears blurred.

The shortest viewing distance of human eye is about 25 cms. At this distance, the limit of resolution is about 0.2 mm. Objects lying together closer than 0.2 mm will not be seen separately by the human eye.

Simple Microscope

The simple microscope is nothing but a single convex lens or a combination of more than one lenses functioning like a single lens. This type of lens when placed between an object and the eye allows the object to come closer, thus forming an enlarged image on the retina. Since the image is larger, the eye is able to resolve finer details that could not be distinguished without the lens.

The magnification obtained with a simple lens can be calculated as follows.

$$M = 250/f \quad + \quad 1$$

Where f is the focal length of the lens in mm, and 250 is the distance of normal vision also in mm.

A single convex lens can be used for magnification upto 3X. Higher magnifications are not possible with a single lens.

A superior quality lens can be constructed by combining two or more lenses into a single unit. Many types of multi-element (unit) lenses have been designed, but even they have limitations. One such multi-element lens is the *triplet* or *aplanat*, which consists of a double convex lens of crown glass sandwiched between a pair of concave lens of flint glass. A magnifying unit of this type can see objects up to 20X. This is commonly used as a field magnifying glass by biologists.

Modern (Compound) microscopes

Modern microscopes, generally called compound microscopes have a complex system of arrangement of lenses that help in higher magnification and better resolution.

Basically the modern microscopes are of two types – light microscopes and electron microscopes. In a light (optical) microscope, the source of illumination of the object is visible light, while in an electron microscope, the source of illumination is a beam of electrons.

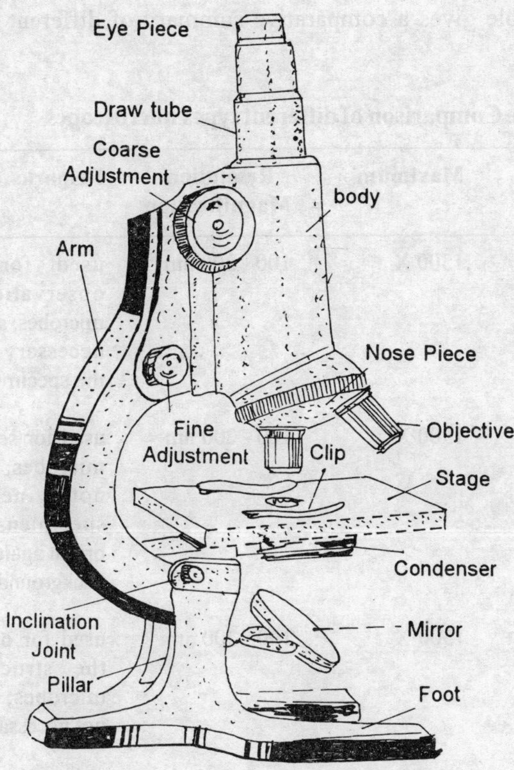

Fig. 3.3 Microscopy
A Compound Microscope

Light microscopes are of the following types –

1. Bright field (Light) microscope
2. Dark field microscope
3. Phase contrast microscope

4. Fluorescence microscope
5. Ultraviolet microscope and
6. Interference microscope

Electron microscopes are of two types.

1. Transmission electron microscope (TEM) and
2. Scanning electron microscope (SEM)

The following table gives a comparative summary of different types of microscopes.

Table Comparison of different types microscopes

No.	Type of Microscope	Maximum	Resolution Magnification	Remarks (uses)
1.	Bright field	1500 X	100 - 200 nm	used for visual observation of microbes; staining is necessary to view the specimens
2.	Dark field	1500 X	100 - 200 nm	used for seeing live microbes; staining not necessary; specimens appear bright against a dark background.
3.	Phase contrast	1500 X	100 - 200 nm	used for observing the structure of microbes; staining not necessary.
4.	Fluorescence	1500 X	100 - 200 nm	useful in the diagnostic identification of microbes; uses fluorescent staining.
5.	Ultraviolet	2500 X	100 nm	better resolution due to the usage of UV (short wave length) light.
6.	Interference	1500 X	100 - 200 nm	used for observation

				of structural details of microbes; produces sharp multicoloured image with three dimensional view.
7.	TEM	500,000 X- 1000000 X	100 - 200 nm	used for the study of ultra structural level of cells and microbes including viruses. Anatomical details of cells can be viewed; highest magnification and resolution possible due to usage of short-wave length electron beam.
8.	SEM	10,000 - 1000000 X	1 - 10 nm	used for the surface viewing of microbes, pollen grains and other objects; surface architecture at the ultra structural level can be seen, three dimensional image is produced.

Bright Field (Light) Microscope

The bright field microscope, or light microscope, or compound microscope is the most common and the most indispensable instrument in the biology laboratory.

Basically the microscope has two types of parts. The mechanical parts provide the structural framework of the microscope and support the optical parts. The optical parts are involved in the magnification of the object. The following are a detailed description of (the various parts) a light microscope.

Base, Pillar and Inclination Joint

The base and the attached pillar serve to support the entire microscope. The base is generally *'V'* (horse shoe) shaped or some times 'U' shaped. Usually the pillar and the base are made heavy in order to minimize vibration. The inclination joint joins the arm of the microscope to the pillar. The joint permits the tilting of the microscope to any degree as desired by the observer.

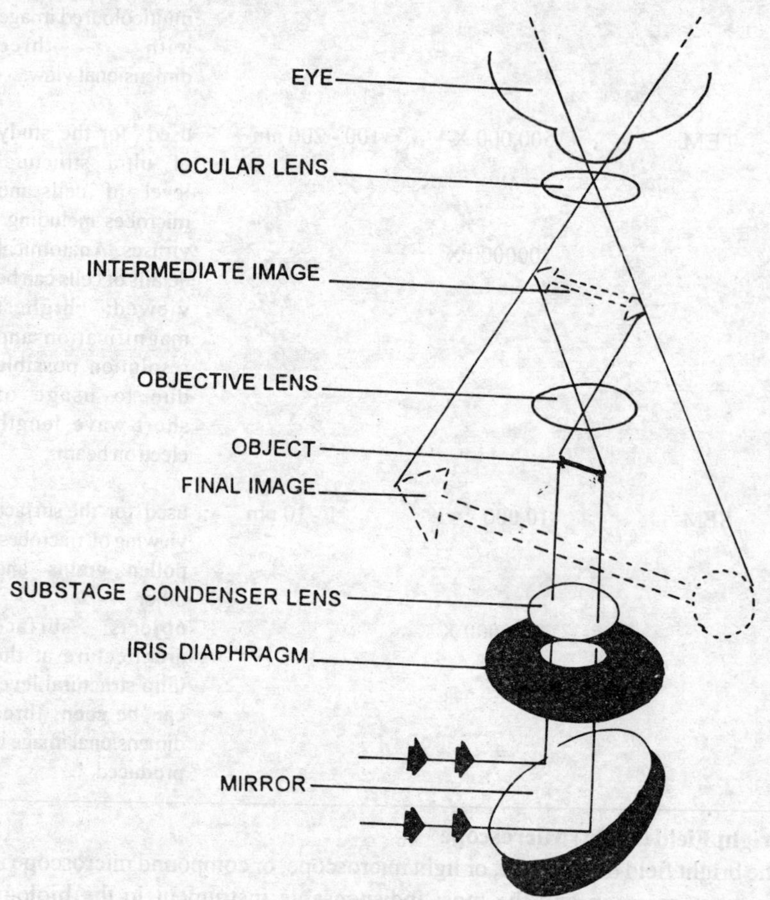

Fig. 3.4 Microscopy
Optical system of the compound microscope

Arm and body tube

The arm is a slightly bent solid piece of iron, which is attached at one end to the pillar by the inclination joint. The other end of the arm has the body tube to which the principal optical systems are attached. The body tube is a cylindrical structure and at the point of attachment to the arm, there is a graduated rack and pinion which helps in the movement of the body tube up and down.

Draw tube, revolving nose piece

Fitting properly inside the upper end of the body tube is the draw tube, which may be drawn up. "The ocular lens fits into the draw tube. The purpose of the draw tube is to adjust tube length so that the objective lens and the ocular are at a specified distance.

The revolving nose piece is a circular rotating disc attached to the body tube. Objective lenses are attached to this.

Adjustment Knobs

The body tube can move up and down on the rack and pinion and the movement is regulated by two sets of knobs – the course adjustment knob and the fine adjustment knob. The coarse adjustment knob brings in greater degree of movement while the fine adjustment knob is used for slight and fine adjustment of the body tube. The adjustment knobs are meant to bring in the object into proper focus. While the range of the coarse adjustment knob is greater, that of the fine adjustment knob is limited.

Stage : This is a small platform attached to the arm of the microscope. This is meant to keep the objects (on sides) for the purpose of observation. It has a hole in the center exactly in line with the body tube and the condenser below to allow the light to pass through. In most modern microscopes, the stage is fitted with a mechanical device called mechanical stage to help in the vertical and horizontal movement of the slide.

Illumination : In order to collect and reflect the light beams there are two kinds of devices attached to the base of the microscope. These are – mirror or an artificial illumination (built in illumination).

The mirror collects the light either from a natural source (sun light), or an artificial source (electric light) and reflects the rays into the microscope. The mirror has two reflective sides– one side is plain, the other side is concave. Usually the concave side is used when natural light is focussed, and the plain side is used while focussing artificial light.

In some of the compound microscopes there is a provision of built in illumination (instead of a mirror), which has a devise to use 8 or 12 volt bulb and provide direct illumination to the stage.

Sub stage Condenser; Iris diaphragm

The light either reflected by the mirror or by a built in source is diffuse and does not have sufficient density. In order to condense the light and focus it on to the object, there is a sub stage condenser. This consists of a condensing lens. The condensing lens may be fixed or in some microscopes it can also move up and down for suitable adjustment on a rack and pinion. The condenser helps to focus all the light on the object. Beneath the condenser there is an *iris diaphragm* which regulates the quantity of light entering into the condenser. The iris diaphragm may be closed and opened with the help of a lever to regulate the entry of light.

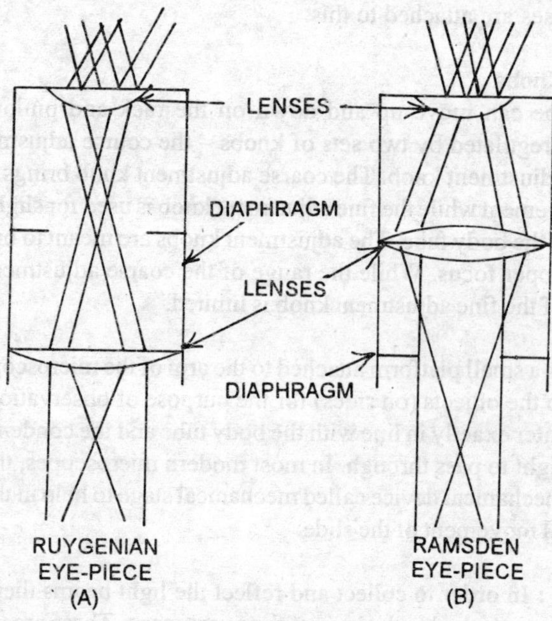

Fig. 3.5 Microscopy
Types of eye-pieces

Optical parts
There are two sets of optics or lenses in a compound microscope. These are eye pieces (ocular) and objectives.

Eye pieces (Ocular) : The ocular or eye piece is a short tube with two lenses which fits into the upper part of the draw tube. The main function of the eye piece is to magnify the image of the object formed by the objective. Eye pieces are marked 5X, 10X, 15X etc.

The eye piece commonly used is the 10X Huygenian eye piece with monocular tube. These eye pieces are also referred to as negative oculars. The positive ocular or Ramsden eye pieces are constructed with convex surfaces of both lenses facing inwards.

Objective lenses : These are attached to the revolving nose piece and are very important as they affect the quality of the image seen by the observer. Based on their type of construction objectives are classified into *achromatic, apochromatic and fluorite.* The first one is simple in construction and less expensive, while the other two are complicated and more expensive and used in costly microscopes. These two are also corrected for most of the common aberrations that occur in the lenses.

The objectives are of the following four types based on their magnification – 10X(Low), 40X(Medium), 60X(High) and 100X(Oil immersion). Several microscope manufacturers use different coloured rings to distinguish individual objectives. They can also be distinguished based on their length; low power being the shortest and oil immersion, the longest.

The oil immersion lens is used for very high magnification. The lens is so named because a drop of a highly refractive liquid (refractive index same as glass) is added on the slide while viewing through an oil immersion objective.

The main functions of the objective lenses are – 1. Concentration of light rays coming from the object, 2. Forming the image of the specimen and 3. Magnifying the image.

Magnification : A microscope is primarily used to enlarge or magnify the object that is being viewed which can not otherwise be seen by the naked eye. Magnification may be defined as the degree of enlargement of an object provided by the microscope. Magnification by a microscope is the product of the individual magnifying ability of the oculars and the objectives. For example if the ocular is 10X, and objective is 40X, the specimen is magnified 400 times. If an oil immersion objective (100X) is used along with 10X ocular, the specimen is magnified 1000 times. The following factors play an important role in magnification.

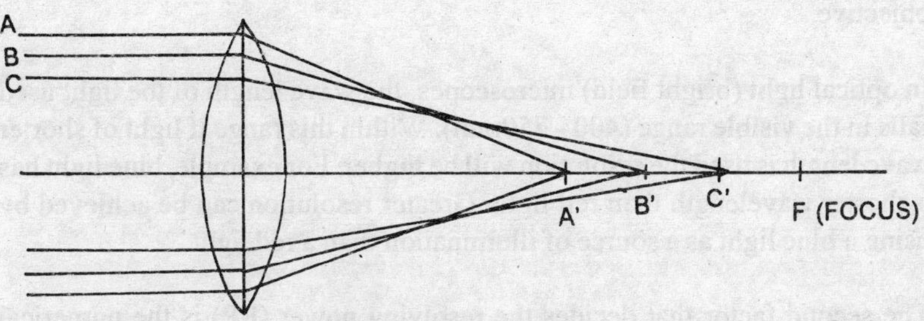

Fig.3.6 Microscopy
Spherical aberration

(i) Optical tube length
(ii) Focal length of the objective lens
(iii) Magnifying ability of the ocular

Total magnification can be calculated as follows:
Total magnification = Optical tube length / Focal length of the objective x magnification of ocular

Resolving power

Theoretically if the magnifying power of the ocular and objectives are increased it should be possible to get higher and higher magnifications. By using high powered lenses a magnification up to 3000 can be obtained but the image will be blurred and details will not be clear. This is due to the fact that in a microscope not only the lenses, but the wave length of the illumination is also important and this decides the resolving power of the microscope.

Resolving power of a microscope is defined as its *ability to distinguish between two particles situated very close.* In a magnified image, the object should not only be larger but the details should also be clear. This is possible when a microscope has the ability to see two points situated very close as two distinct entities. In other words, resolving power may be said to be the minimum distance at which two structural entities of an object can be visualized as discrete individual structures even in the magnified image.

(The above explanation may be illustrated clearly with a comparison with the human eye. Human eye functions on the same principal as that of a light microscope i.e., we see objects because of the light reflected by them. The human eye has a resolving power about 0.25 mm in the sense, two dots situated 0:25 mm (or more) apart can be seen as two dots; any thing closer will look like a single dot).

The resolving power of a microscope depends on two factors – the wavelength of the light used for illumination and the numerical aperture (NA) of the objective.

In optical light (bright field) microscopes, the wave length of the light used falls in the visible range (400 - 750mm). Within this range if light of shorter wave length is used the resolution will be higher. For example, blue light has a shorter wavelength than red light. Greater resolution can be achieved by using a blue light as a source of illumination than a red light.

The second factor that decides the resolving power (RP) is the numerical aperture of the objective. NA is defined as the property of a lens that decides the quantity of light that can enter intc it. It depends on two factors.

(i) Refractive index of the medium that fills the space between the specimen and the front of the objective lens and

(ii) The angle of the most oblique rays that can enter the objective lens (The more divergent or oblique rays that an objective can admit, greater is the resolving power).

NA can be mathematically calculated by the following formula

NA = *n* sin *f* Where

n = refractive index of the medium

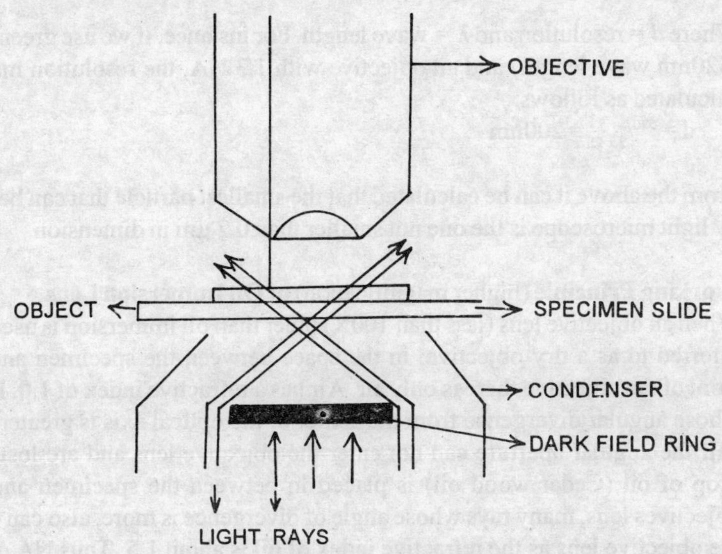

Fig. 3.7 Microscopy
Dark field microscope

f = angular aperture. This is defined as the angle between the most divergent rays passing through the lens and optical axis of the lens. Rays cannot enter the objective if their angle of divergence from the normal rays (straight rays) is greater than half the angular aperture.

Using NA, the resolution power of the microscope can be calculated as follows.

RP = Wave length of light / 2 x NA

For instance, if a yellow light of wave length 580nm with NA of 1.0 is used, RP will be as follows:

RP = 580/2x1 = 290nm

If a blue light is used the RP will be still higher (i.e. it can resolve objects smaller than 290nm).

Limit of resolution : This may be defined as the shortest distance between two objects when they can still be distinguished as two separate entities. Highest resolution in a light microscope is possible with the shortest wave length light (visible), and with the objective having the highest NA. This can be expressed as follows:

$$d = \lambda / 2NA$$

Where d = resolution and λ = wave length. For instance, if we use green light (520nm wave length) and an objective with 1.3 NA, the resolution may be calculated as follows:

$$d = {}^{520}/_{2 \times 13} = 200\text{nm}$$

From the above it can be calculated that the smallest particle that can be seen by light microscope is the one not smaller than 0.2 µm in dimension.

Working Principle (higher magnification) **of Oil Immersion Lens**

When an objective lens (less than 100X) other than oil immersion is used it is referred to as a dry objective; in the space between the specimen and the front of the objective there is only air. Air has a refractive index of 1.0. Rays, whose angular divergence from the center of the optical axis is greater than half the angular aperture can not enter the objective lens and are lost. If a drop of oil (Cedar wood oil) is placed in between the specimen and the objectives lens, many rays whose angle of divergence is more, also can enter the objective lens as the refractive index of oil is about 1.5. Thus NA of the objective is increased the which results in better resolution and higher magnification.

Types of Condensers : Three types of condensers are in use in microscopes. These are –

***Abbe's Condenser* :** This is the simplest one fitted to a microscope. There are two lenses in this, but both are not corrected for any chromatic or spherical aberration. This type of condenser is good only for low power viewing. At higher magnifications, the images of the specimens get blurred due to improper focussing.

***Aplanatic Condenser* :** In this, there is a third lens added to the original two, to correct the spherical aberration only and not just chromatic. Aplanatic condenser gives a good image when used with monochromatic light.

Achromatic Condenser : This focuses most of the incident light on the object and allows very little light to escape and produce glare. The NA of the condenser generally equals that of the objective. In this condenser, both spherical and chromatic aberrations are corrected.

Aberrations in the objectives

The surfaces of objective lenses are spherical and such surfaces can not form perfect images. This is due to variations in refraction and focus by the peripheral parts of the objectives. These cause defects or aberrations in the images. Aberrations in objectives are of two types - *Spherical aberrations and Chromatic aberrations.*

Spherical aberrations result due to passing of lenses–which cannot be brought to the same focus as those which pass through near the optical axis. As a result the image gets distorted.

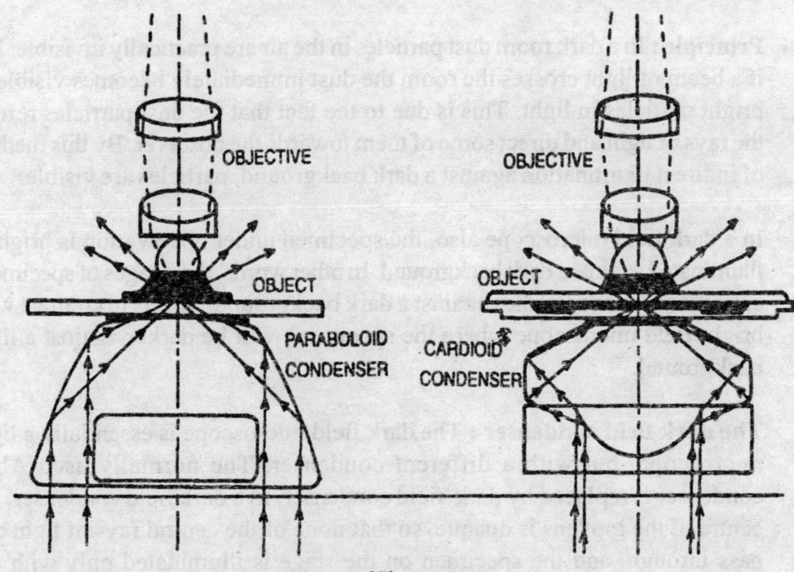

Fig. 3.8 Microscopy
Two types of dark field condensors

Chromatic aberration occurs due to the splitting of light into its individual component colours as it would happen when it passes through a prism. These monochromatic beams of various wavelengths are not recombined to the same focus due to varying refraction of lens. The result is a hazy image.

Correction of aberrations : Aberrations may be corrected by using better quality lenses, or by the combination of lenses with varying dispersive qualities. A lens system corrected for chromatic aberration of two colours is called *achromatic* and the one corrected for three different colours is called *apochromatic.* Crown glass with low dispersive power is used for convex lenses, while flint glass with high dispersive power is used for concave lenses. A combination of the two will result in an achromatic lens system.

Spherical aberrations are corrected by the proper use of iris diaphragm to keep out peripheral rays. Additionally, the lens may be ground to a shape which will permit the peripheral and axial rays to have a common focus.

DARK FIELD MICROSCOPE

There are several alternative types of microscopes other than bright field microscope that increase the contrast between the background and the specimen without staining, thus allowing for visualization of living specimens. One such microscope is dark field microscope.

Principle : In a dark room dust particles in the air are practically invisible. But if a beam of light crosses the room the dust immediately becomes visible as bright particles in light. This is due to the fact that the dust particles refract the rays of light and direct some of them towards the observer. By this method of indirect illumination against a dark background, particles are visible.

In a dark field microscope also, the specimen under observation is brightly illuminated against a dark background. In other words, the images of specimens appear as luminary bodies against a dark background. This is in contrast with bright field microscope where the specimens will be darker against a light background.

The dark field condenser : The dark field microscope is essentially a light microscope, but with a different condenser. The normally used Abbe condenser is replaced by dark field condenser. In a dark field condenser, the centre of the top lens is opaque, so that none of the central rays of light can pass through, and the specimen on the stage is illuminated only with the peripheral oblique rays. The light rays are focussed in such a way that they do not reach the specimen directly (as in the ordinary microscope) but instead the light rays traverse across the specimen horizontally almost at right angle to the objective lens. As a result of this the field appears dark, but the specimen stands out as a bright refractive structure in the same way as a dust particle in a beam stands out due to its shining.

In normal operations of dark field microscopy, a modified version of Abbe condenser with low power objectives would suffice. But for higher magnifications, special dark field condensers with cardoid and paraboloid lenses are employed. In some instances ordinary light may not be enough. Sources of intense light such as *carbon arc* lamps may become necessary. Utmost care should be taken to use perfectly clean and scratch free glass slides and cover slips. Many a time unwanted particles or dust are capable of reflecting light and may illuminate the background which should always remain dark.

Uses of dark field microscope : The dark field microscope is used for examination of live, unstained preparations of microbes or other specimens suspended in fluid. It is highly useful for the study of very small bacteria which are generally invisible under ordinary microscope. Motility of microbes also can be observed since they (microbes) are in live condition and are not killed.

PHASE CONTRAST MICROSCOPE

In bright field microscopes, transparent protoplasmic contents cannot be differentiated unless they are stained. This is due to the fact that when light rays pass through both medium and the specimen, there is very little difference or contrast in their refractive indices. As a result, the alteration in the rays that pass through the specimen or the medium will be almost the same when viewed under a light microscope. In order to increase the contrast between the specimen and the background or to bring about differences in the refraction of the rays that pass through the specimen and those that pass through the medium, dyes (stains) are used that selectively stain different cell organelles. Staining, however, leads to cellular death and when we observe a stained specimen we will not be seeing a live specimen.

Phase contrast microscope is a special microscope that enables the viewer to see transparent protoplasmic components without staining and without killing. Originally developed by Frederick Zernike (1933), hence called *Zernike microscope,* Phase contrast microscope is the ideal instrument for observation of living protozoans and other transparent microbes, without staining. Frederick Zerinke was awarded–Nobel Prize in physics in the year 1953 for the discovery of phase·contrast principle.

Principle : In order to understand the principle behind the working of a phase contrast microscope we must first understand how image contrasts are produced and the variable behaviour of the light rays as they pass through an object.

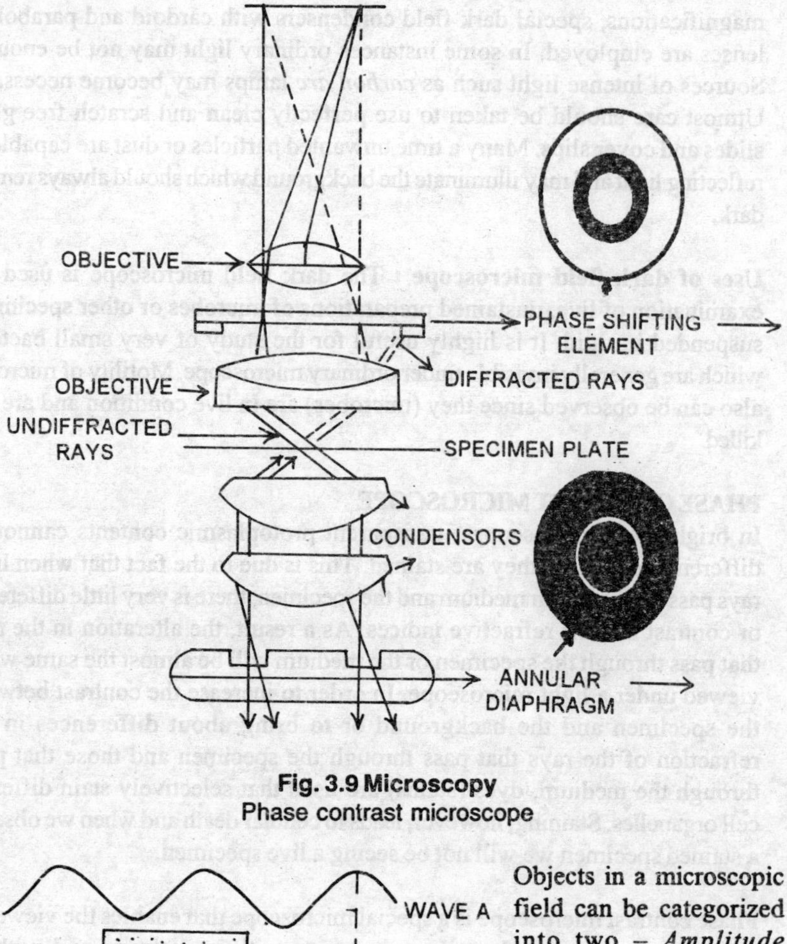

OBJECTIVE
PHASE SHIFTING ELEMENT
OBJECTIVE
DIFFRACTED RAYS
UNDIFFRACTED RAYS
SPECIMEN PLATE
CONDENSORS
ANNULAR DIAPHRAGM

Fig. 3.9 Microscopy
Phase contrast microscope

WAVE A
WAVE B
WAVE C
λ/4
λ/2

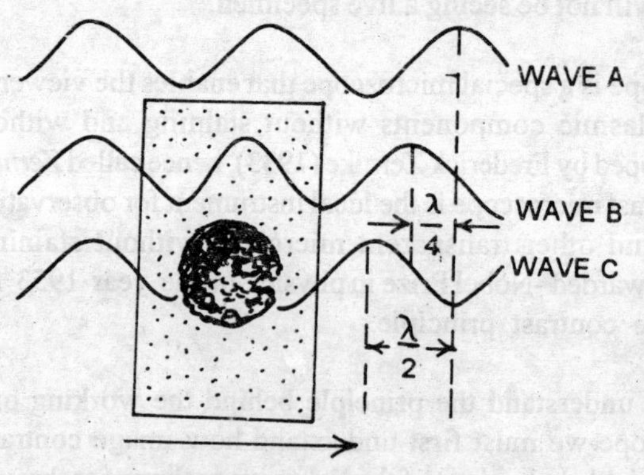

Fig. 3.10 Microscopy
The retardation of phases of light waves

Objects in a microscopic field can be categorized into two – *Amplitude objects* and *Phase objects.* Amplitude objects show up as dark objects under the microscopic view because of the reduction in the intensity (amplitude) of the rays that pass through. Phase objects which are transparent on the other hand allow the light rays to pass through without any reduction in their intensity. But when the light rays are passing through a transparent object, some rays will have retardation by about one-quarter wave length. This retardation is called *phase shift.* But the phase shift will not cause any change in amplitude; thus the objects appear transparent. The 1/4 wave length phase shift is utilized in the phase contrast microscope to create image contrast.In order to understand this, first we should understand what happens to the rays when they pass through a transparent object.

Behaviour of rays : Light rays passing through a transparent object emerge out in two different ways – *direct* and *diffracted.* The rays that pass through the object in a straight line are called direct rays and are unaltered in amplitude and phase. The rays that are bent and slowed down as they pass through due to differences in density of the medium, are called diffracted rays. It is this relationship between the rays that is made use of in the phase contrast microscope.

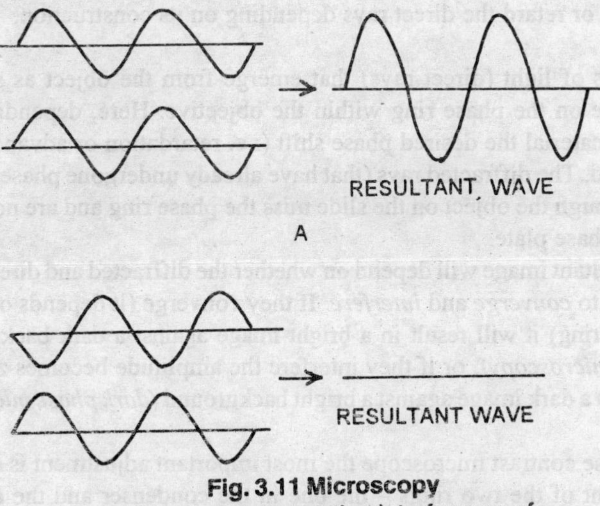

Fig. 3.11 Microscopy
Additive and destructive Interference of light waves

If the direct rays and diffracted rays of an object can be brought into the same phase (the crest of both light waves coincide) with each other, the resultant increase in amplitude due to the sum of both the converged rays and is called *coincidence.* As a result of the increased amplitude, the object looks very bright in the field. On the other hand, if the direct and diffracted rays are out of phase (i.e. the crest of one wave coincides *trough* but and *not with the crest* of another wave) it is said to be *reverse phase* and their amplitudes cancel each other and the object looks dark. This is called *interface.* Both these phenomena (coincidence and interference) are used in a phase contrast microscope.

The phase contrast set up

The basic construction of a phase contrast microscope is like a bright field microscope only except for two special attachments. These are – (i) a special type condenser and (ii) a phase plate.

The condenser has a special diaphragm consisting of an *annular stop.* This annular stop can be compared to a solid rod kept loosely at the centre of a cylinder. The annular stop allows only a hollow cone of light rays to pass through the condenser and light the object on the slide.

The phase plate is a special optical disc located in the rear focal plane of the objective. It has a special phase ring coated with a material that can either advance or retard the direct rays depending on its construction.

The rays of light (direct rays) that emerge from the object as solid lines converge on the phase ring within the objective. Here, depending on the coated material the desired phase shift (i.e. retardation or advancement) is produced. The diffracted rays (that have already undergone phase shift) that pass through the object on the slide miss the phase ring and are not affected by the phase plate.

The resultant image will depend on whether the diffracted and direct rays are allowed to *converge* and *interfere.* If they converge (it depends on the type of phase ring) it will result in a bright image against a dark backgrounding *(phase microscopy)*, or if they interfere the amplitude becomes zero and it results in a dark image against a bright background *(dark phase microscopy).*

In a phase contrast microscope the most important adjustment is the proper alignment of the two rings – the one in the condenser and the one in the objective. The ring of direct light that comes out of the object must correspond to the phase ring in the objective.

How an unstained preparation can be seen under a phase contrast microscope? In any specimen or cell there will be, difference in thickness between the structures of components. When the light rays pass through these there will be variable refraction of the rays and these phase changes are converted into visible differences of light intensity (due to the phase ring). Even some structures which are difficult to stain and observe under light microscope become conspicuous in a phase contrast microscope, because small phase changes result in interference of light waves resulting in high contrast images.

FLOURESCENCE MICROSCOPE

Flourescence microscope involves staining of the specimens with special flourescent dyes and is an indispensible instrument in diagnostic microbiological laboratories.

A flourescence microscope differs from an ordinary light microscope in the following respects.

(i) Light source is a mercury vapour lamp

(ii) A dark field condenser is used instead of the normal Abbe condenser

(iii) Three sets of filters are employed to alter the light rays that pass through the instrument and reach the eye.

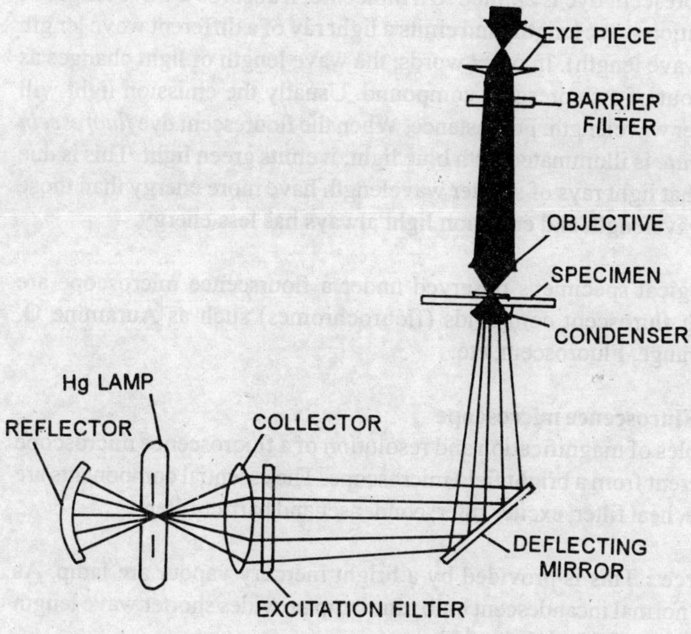

Fig. 3.12 Microscopy

Fluorescence microscope diagrammatic representation of the path of rays and action of excitation and barrier filters

The Principle of flourescence

Light exists as a form of energy and is transmitted in the form of waves. When a wave of light encounters any obstruction like when it reaches molecules of a substances, the electrons of the molecule at the outer orbit get excited and oscillate in resonance with the light waves. When a substance absorbs light it essentially absorbs energy. This energy that is absorbed can not just disappear. It should be transmitted out in some form or it should be converted to some other form of energy as it happens in photosynthesis. When light is transmitted by the energized (excited) molecules it is said to be *Photoluminiscence.* In *photoluminiscence*, the light absorbed is not transmitted immediately. There is always a time lapse between absorption and emission. If the time lapse is greater than 10 (1/10,000) seconds it is said to be *Photoluminiscence.* If on the other hand the time lapse is less than 10 seconds it is known as *fluorescence.*

When a fluorescent dye is applied to a molecule, it absorbs a wave length of light (excitation wave length) and emits a light ray of a different wave length (emission wave length). In other words, the wave length of light changes as it emerges out of a fluorescent compound. Usually the emission light will have a longer wavelength. For instance, When the flourescent dye *fluorescein isothiocyanate* is illuminated with blue light, it emits green light. This is due to the fact that light rays of shorter wavelength have more energy than those of longer wavelength and emission light always has less energy.

Microbiological specimens observed under a flourscence microscope are coated with fluroscent componds (flourochromes) such as Auramine O, Acridine orange, Fluoroscein, etc.

Parts of a Fluroscence microscope

The principles of magnification and resolution of a fluoroscence microscope are no different from a bright field microscope. The essential components are light source, heat filter, exciter filter, condenser and barrier filter.

Light source : This is provided by a bright mercury vapour arc lamp. As against the normal incandescent bulb, this lamp provides shorter wave length light rays such as UV, violet and blue.

The lamp always comes with a transformer as sometimes a voltage of 18,0000 volts are required.

Mercury vapour lamps produce light rays in range of 200 - 400nm (UV) and visible rays in the range of above 780nm.

Mercury vapour lamps are not only expensive but also harmful. Hence several precautions are necessary. The bulbs are pressurized, hence there is danger of their explosion. Eyes should not be directly exposed to the bulb as the rays are harmful.

Heat filter : Infrared rays produced by the lamp generate considerable heat, besides they are of no use in fluoroscence. In order to eliminate these rays a heat filter is placed in front of the lamp and before the condenser to absorb heat. Heat filter, however, does not prevent the transmission of UV and the visible rays.

Exciter filter : The light cooled down by the heat filter, next passes through the exciter filter which absorbs all but shorter waves that are needed to excite the flurochrome dye coated specimen on the slide. These filters which are dark, allow only green, blue, violet or UV rays.

Condenser : For best results, always a dark field condenser is used, because in a dark background even mild fluoroscence can be easily detected. Another advantage of this condenser is that it deflects majority of the UV rays thus protecting the observer's eyes. In order to achieve this the NA of the objective is always kept at 0.05 less than that of the condenser.

Barrier filter : This is situated in the body tube of the microscope between the objectives and the eye piece to remove all remnants of the exciting light so that only the fluoroscence is seen. When excitation is by UV, the exciter filters are dark and the barrier filters will be almost colourless. On the other hand, if blue excitation is used, the barrier filters are either yellow or deep orange in colour.

Hints for using the microscope : The following points are to be kept in mind while using a fluorescent microscope for best results.

1. *Warm up :* Fluorescent micrscopes require a warm up period. The illumination is low when turned on but reaches the maximum in about two minutes. Optimum illumination is obtained in about 30 minutes. It is better to switch on the light 45 minutes before using.

2. *Check of mercury lamp :* The life expectancy of a mercury arc lamp is about 400 hours. Hence it is necessary that a log should be maintained regarding the number of hours that the instrument is used. A check should be made after approximately 200 hours of usage.

3. *Selection of filter:* Proper combination of exciter and barrier filters are necessary to achieve best results. The most frequently used filters are –

Bluish Schott BG12 - Exciter filter

Yellowish Schott OG I Barrier filter

Examination : For preliminary exmination like locating a particular part or cell of the specimen, it is better to use ordinary illumination and use the mercury arc illumination only for precise observation. If the specimen has to be observed under oil immersion, special low fluorsecing immersion oil is to be used. Ordinary immersion oil used for bright field microscopy should not be used.

Safety measures : Fluorescent microscope can be harmful and injurious if not used properly. The following precautions are necessary to avoid any harm to the user.

1. The pressurized mercury arc bulb may explode under certain circumstances. But the equipment is designed to prevent this. However the lamp should not be inspected when it is hot. The lamp is safe when relatively cold. Usually 15 - 20 minutes is sufficient to cool the lamps.
2. Do not expose your eyes directly to the mercury lamp. The equipment is designed in such a way as to prevent the scattering of the rays. Please note, that the unfiltered rays from the lamp contain UV as well as infrared rays both of which are harmful to the eyes.
3. Make sure that both the filters (of the suitable type), are in their place before looking into the microscope. Removal of any one or both is harmful to the to the eyes.

ULTRAVIOLET MICROSCOPE

In Ultraviolet microscope, the optics are specially meant to transmit or reflect UV light (200 - 350nm). As the source of illumination (UV) and having a shorter wave length, UV microscopes have a better resolving power than light microsopes. It is not possible to look into a UV microscope directly as the rays are injurious. Hence, the images of the object are either photographed or projected on using UV sensitive TV cameras.

INTERFERENCE MICROSCOPE (Stereomicroscope)

The interference is essentially based on the same principle as that of phase contrast microscope. In both the microscopes, the principle of increased or decreased amplitude due to the interference between *out of phase* light waves and *in phase* light waves is used to produce contrast in the image. While a single beam of light passing through the object is made use of in phase contrast microscope, there are two beams of light that combine after passing through the specimen. This produces an interference pattern. Interference microscopes are more versatile in comparison with phase contrast microscopes, as they have objectives of higher NA and also use colour to increase the contrast.

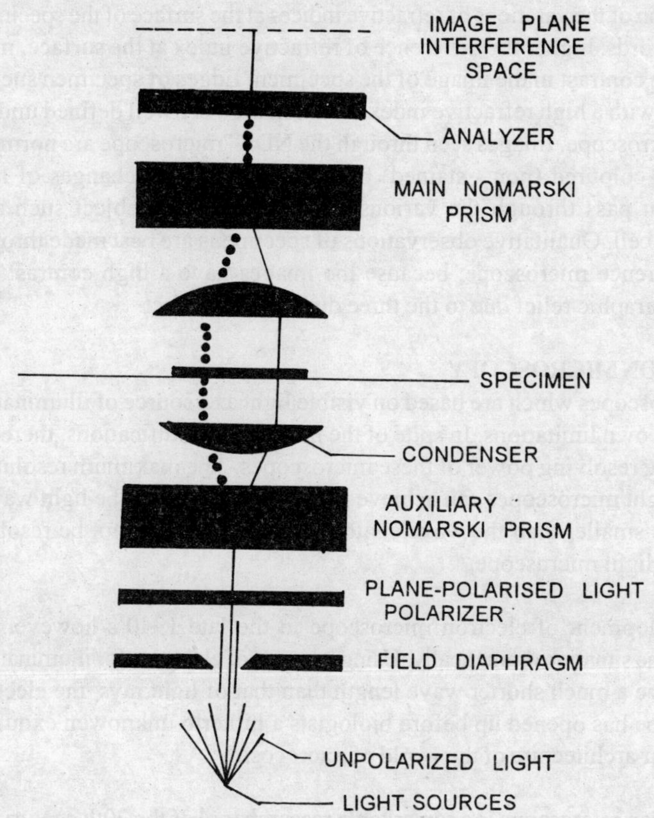

Fig. 3.13 Microscopy
The Nomarksi Interference microscope

One of the most frequently used interference microscope is a *Nomarski Differential Interference Contrast(NDIC) Microscope* which can produce a high contrast, almost three dimensional image of transparent objects. The special features of a NDIC microscope are a polarizing filter, and interference contrast condenser and a prism analyzer plate.

In the NDIC microscope a beam of polarized light split into two beams at right angles to each other; are made to travel through the specimen and are combined and forced through an analyzer. The resultant interference pattern produces a pseudo three dimensional image. This is due to the fact that the two beams of light passing through the specimen produce a stereoscopic effect. The degree or level of three dimensional configuration that is visualized is a function of the varations of refractive indices at the surface of the specimen. In other words, higher the difference of refractive index at the surface, more will be the contrast in the image of the specimen. Edges of specimen such as cell walls with a high refractive index variation are very well defined under a NDIC microscope. Images seen through the NDIC microscope are normally brilliantly coloured (non - stained) because of the phase changes of light waves that pass through the various components of an object such as a microbial cell. Qualitative observations of specimens are best made through an interference microscope, because the images have a high contrast and high topographic relief due to the three dimensional effect.

ELECTRON MICROSCOPY

The microscopes which are based on visible light as a source of illumination have their own limitations. In spite of the numerous modifications, there is a limit to the resolving power of these microscopes. The maximum resolution that the light microscopes can achieve is the wave length of the light waves. A particle smaller than the wavelength of visible light can not be resolved under the light microscope.

The development of electron microscope in the late 1940's however has changed the situation dramatically. Using a source of electrons for illumination, which have a much shorter wave length than that of light rays, the electron microscope has opened up before biologists a hitherto unknowen exquisite subcellular architecture of the world of microcosm.

The electron microscope is a remarkable research tool of the 20th century. Its power of magnification runs into 100000X or a resolving power of 10^{-9} centimeters. Experiments that eventually lead to the development of electron microscope began as early as 1920. The first prototype model was developed for physical applications.

Principle of working

As the name indicates, in an electron microscope, a beam of electrons is used as a source of illumination, instead of a beam of light as in light microscopes. In order to understand the working mechanism of an electron microscope, we must first understand some elementary facts about electrons and their behaviour.

Electrons are negatively charged sub atomic particles having a mass of 9.1 x 10^{-28} grams. An integral part of all atoms, electrons orbit round the atomic nucleus at velocities far exceeding 50,000 kms per second. When the atoms of a metal are excited by heat energy to an extent where the electron velocity exceeds the attraction of the nucleus, they (electrons) fly off from the clutches of the nucleus and are lost. When a metal, like tungsten is heated by applying a high voltage current, electrons come out in a continuous stream and this can be directed to form a high velocity electron beam.

Louis de Bragile (1924), working on electrons opined that electrons being particulate must behave like light waves and should have both a fixed wave length and frequency. He used the following formula to calculate the wave length λ (lambda) of the matter.

$$\lambda = h / mv$$

where h is planck's constant, m is mass and v is velocity. Calculations based on this equation have shown that electrons accelerated to a potential of 60,000 volts have a wave length of 0.05 Å (Angstorm) or 105 times smaller than that of visible light. Hence an electron beam should be able to resolve (theroretically) particles 105 times smaller than that of the wave length of light.

The discovery that electrons have wave motion and they can be focussed on to an object in the form of a beam (like light ray) in a magnetic field provided the impetus for the development of electron microscope. By the late 1930's, a system of magnetic coils capable of focussing an electron beam had been designed. It was also shown, that by regulating the current flowing through the magnetic coils, the magnification can be regulated. Theoretically, an enhancement of resolution power equivalent to the wave length of a beam of electrons should be possible in an electron microscope.

But, there are practical limitations in the design of magnetic coils (equivalent to lenses in a light microscope) used for focussing the beam of electrons. The main problem is the one of distortion produced when the angle of illumination is more than a few tenths of a degree. As a result, to catch the beam of electrons, the numerical aperture of objectives has to be much less than that of light microscopes. The actual resolving power of electron micrtoscopes is only about 0.05nm; far less than the theoretical potential of 0.002nm. But even then, 0.05nm is much much smaller when compared with 100nm, that is the limit in a light microscope.

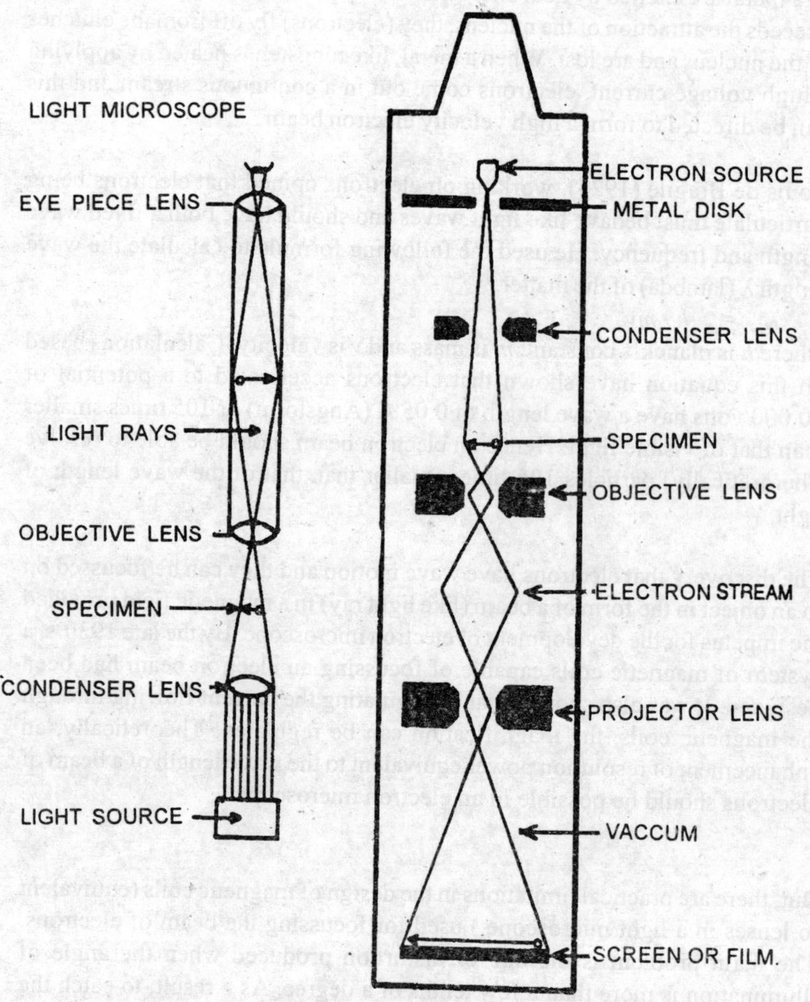

Fig. 3.14 Microscopy
Light and Electron Microscopes (comparison of paths of light)

Basic construction of an electron microscope

The modern electron microscopes have a common basic construction pattern with standard components such as

(i) Electron generator (gun)
(ii) Electromagnetic coils (lenses) - 3 sets
(iii) Screen (for viewing) and
(iv) Vacuum pump

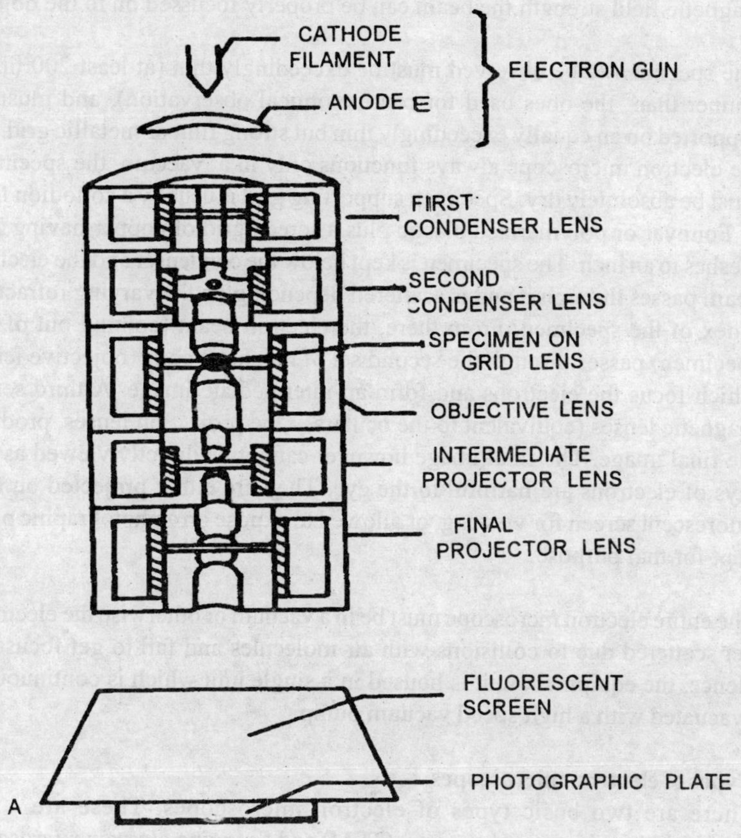

Fig. 3.15 Microscopy
The transmission electron microscope

The electron generator is the source of illumination. Also called electron gun, it is a tungsten filament, which when heated by electric current emits a stream of electrons. The electrons accelerated by the high voltage energy are forced through a *collimating* aperture which renders the rays in parallel lines and fashions them into a beam. From there, the beam passes through the condenser. The condenser 'lens' is a magnetic coil which corrects the aberrations (bending) in the beam and directs them towards the object. The strength of the magnetic lens depends on the

amount of current that is allowed to flow through it. Greater the flow of current, greater will be the strength of the magnetic field and greater will be the bending effect on the rays of electron. In other words, by varying the magnetic field strength,the beam can be properly focussed on to the object.

The specimen to be observed must be exceedingly thin (at least 200 times thinner than the ones used for routine optical observation), and must be supported on an equally exceedingly thin but strong film or metallic grid. As the electron microscope always functions only in a vaccum, the specimen must be absolutely dry. Specimen supporting grid is usually a collodion film or Fonnvar or polymerized plastic plus a screen grid of copper having 200 meshes to an inch. The specimen is kept below the condenser and the electron beam passes through it and is scattered depending on the varying refractive index of the specimen. From there, the electron beam (coming out of the specimen) passes through the second set of magnetic coils (objective lens), which focus the electrons and form an intermediate image. A third set of magnetic lenses (equivalent to the oculars) - the projection lenses, produce the final image. The final image however cannot be directly viewed as the rays of electrons are harmful to the eye. They are either projected on to a fluorescent screen for viewing, or allowed to expose on a photographic plate kept for that purpose.

The entire electron mcroscope must be in a vacuum or otherwise the electrons get scattered due to collisions with air molecules and fail to get focussed. Hence, the equipment unit is housed in a single unit which is continuously evacuated with a high speed vacuum pump.

Types of electron microscopes

There are two basic types of electron microscopes. These are – the *Transmission electron microscope* (TEM) and *Scanning electron microscope* (SEM).

Transmission Electron Microscope (TEM)

In a TEM, the beam of electrons passes through the specimen forming an image of the detailed structure of the specimen. This is similar to the light microscope in construction except for

(i) source of illumination is a beam of electrons (not visible light)

(ii) There are a series (usually three sets) of electromagnets for focussing the beam of electrons. These are called magnetic lenses (not made out of glass).

The general construction pattern of TEM is the same as explained above. TEM is for sstudying the structural details of the various components present in a cell or microbe.

The specimens form a contrasting image with all the structural details.

Depending on their refractive index, the components of a specimen bring about varying degrees of scattering of the electron rays. Generally, however, biological specimens do not produce a great degree of scattering of electrons,hence the contrast is low. Staining is resorted to improve the contrast. But instead of dyes used for staining in light microscopy. electron dense heavy metal salts are used for staining in electron microscopy.

Some problems are encountered while viewing a specimen through TEM. Many a time artifacts appear which can be easily mistaken to be a component of a cell. (An *artifact* is the appearance of something in an image due to causes within the optical system, or due to the improper preparation of a specimen and does not represent any component of the specimen under view). Proper preparation of the specimen, correct adjustment of the electron beams, accurate vaccumization will help in reducing the chances of artifact appearance.

Preparation of specimen for TEM

The biological specimens to be examined have to be specially prepared in order to obtain proper structural details with high magnification, but at the same time avoiding the appearance of artifacts. The following steps are necessary before the specimens are ready for observation under TEM.

Dehydration and Fixation : Under light microscopy, specimens are kept in water or oilier liquids for observation. But in an electron microscope, the specimens have to be absolutely dry. If any water is present, the specimen would boil (as it is in a vacuum) resulting in disintegrating the structural organization of the specimen. Additionally, the specimen also has to be fixed in its proper orientation. Fixation and dehydration have to be carried out in several stages in order to prevent distortion.

Ultra Sectioning : From the point of view of magnification obtained in an electron microscope, microbes are too thick for viewing. Hence, normally they are cut into thin sections using an ultra microtome. The ultra microtome has a device which advances the fixed specimen on a diamond or glass knife surface at a known thickness so that sections of uniform thickness are cut. The microbes are usually embedded in a plastic resin to facilitate sectioning.

Staining : In order to enhance the contrast, the specimens are stained with heavy metal containing compounds such as phosphotungstic acid. The staining can be negative or positive. In *negative staining,* the sectioned specimen is dipped in heavy metal solution which stains the background and not the specimen. The specimens are then visualized in relief against a dark background.

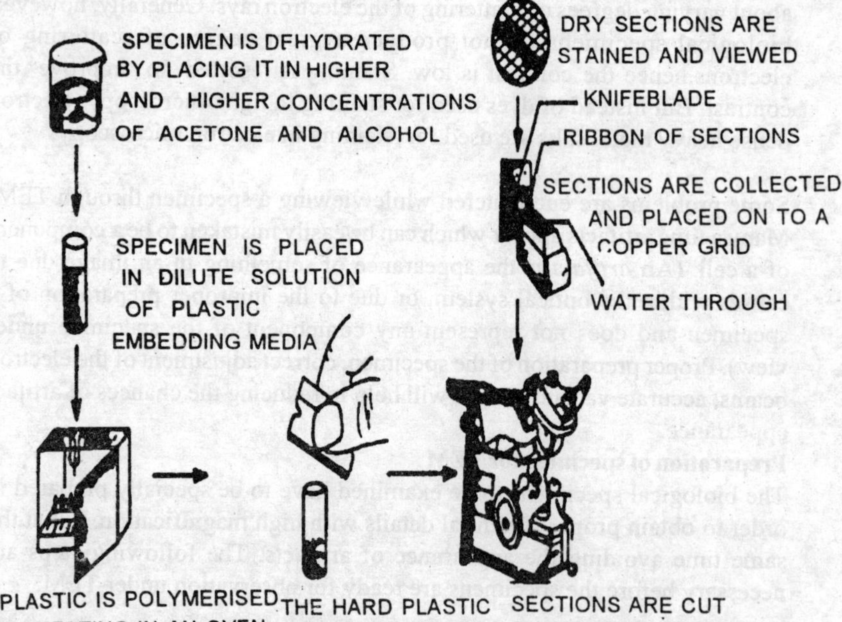

Fig. 3.16 Microscopy
Preparation of specimen for viewing through TEM

Freeze etching **:** This is a special technique employed mainly to reveal the details of structure in microbes. In freeze etching, various biochemically defined layers are made visible, including the organelle details. In this technique (instead of using fixatives to preserve the cell structure) the specimens are rapidly struck with a knife blade. At this temperature (freezing), biological specimens are hard to cut, but they crack along the lines of their natural weakness i.e., they are fractured. The fractured specimen is then *etched* i.e., the water (ice) is allowed to evaporate from the surface. This evaporation raises the surface layers of the specimen. A replica, then is made of the freeze etched specimen, by exposing it to vapours of heavy metal (platinum) at 45^0 angle to produce a shadow effect. The specimen then is rotated at 90, and exposed to vapourized carbon. This produces a replica of the surface of the specimen. After removing the remmants of biological material, the carbon replica is viewed under electron microscope. Surface details (internal as well external) of organelles can be clearly visualized by the freeze etching technique.

Scanning Electron Microscope (SEM)

SEM is primarily used for visualizing the surface architecture of the specimen (pollen grains, hairs, membranes, etc.) rather than the internal details. In a SEM, an electron beam is scanned across the surface and a three dimensional image is produced.

The construction plan and working principle of a SEM is different from that of TEM. In SEM, an accelerated beam of electrons is produced from the electron gun and is focussed on the specimen by the condenser lens. The magnetic lenses of a SEM are so constructed as to produce an extremely thin beam of electrons.

The primary electron beam as it strikes the specimen, forces, out electrons from the surface (of the specimen). These are the secondary electrons and are transmitted to a collector. During this process, some of the primary electrons also are reflected and transmitted to the collector, but their (primary electrons) number is far less than the secondary electrons. As a result, the image signal is developed more by the secondary electrons than by the primary electrons. The electrons are then transmitted from the collector to a detector which has a substance that emits light when struck by electrons. The light so emitted is converted to an electrical current which is used to control the brightness of an image on a CRT (Cathode Ray Tube) screen.

The secondary electrons deflected out of the specimen will be a replica of the refractive index of the surface and thus produce an image on the CRT screen revealing all the topographical details. Image contrast mainly depends on surface topography which determines the number of secondary electrons reaching the detector. The image on the CRT screen will be three dimensional.

Magnification of the image of the specimen in SEM is not achieved through the lenses as in a light microscope or TEM. It is dependent upon the ratio of the length of the scan across the specimen surface to the length of the scan of CRT. For instance if the electron beam scans 100 nm across of specimen and the image on CRT is 100 nm, the magnification is 100000 times. Thus, one can decide the magnification (depending on the requirement) by adjusting the scan distance across the specimen.

SEM also has a resolution equal to that of TEM. A resolution from 1-10 nm is possible with a corresponding magnification from 10000-100000.

Preparation of specimen for SEM

1. *Dehydration :* As SEM also operates on a vacuum like TEM, total dehydration is necessary, but as surface topography is essential, dehydration

should be done in such a way as not to disturb the surface configuration. This is achieved by critical point drying which minimizes artifact formation. In critical point drying, at a particular temperature and pressure, the liquid changes to gas without any surface tension damage to the specimen. The specimen is first immersed in ethanol or acetone to remove water and then in pressurized liquid of CO_2 simultaneously raising the temperature above 32ºC, the critical point of CO_2 . At this temperature range, the liquid vapourizes without surface tension leaving the specimen perfectly dry.

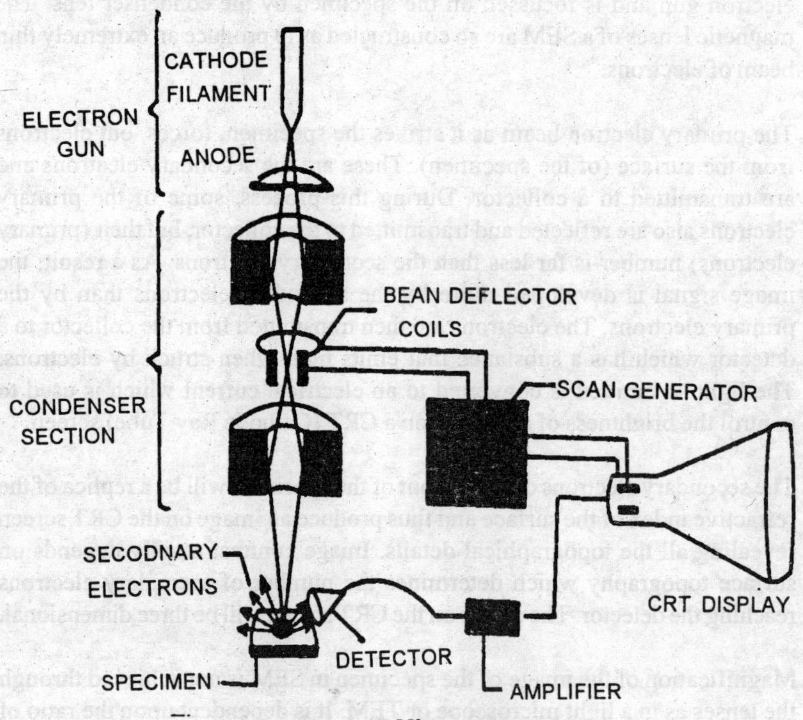

Fig. 3.17 Microscopy
The scanning electron microscope

2. *Shadow casting :* In this technique, the specimen is coated with an extremely thin layer of gold, gold-palladium or platinum at an oblique angle, so that the specimen produces a shadow on the uncoated side. The shadow technique results in the production of a three dimensional topographic image of the specimen. Coating is done with a device called, *a sputter coater.* In some cases, double shadowing also can be done by coating the materials from two different angles.
3. *Surface replica:* In this technique, widely used for the study of

architectural pattern of the wall surface of spores, pollen etc, a thin layer of a coherent material is coated on to the specimen which is then floated on to a water surface, from where it is transferred on to a strong acid or alkali. This dissolves the specimen without damaging the replica. The replica is then dried and kept on the metal grid for viewing.

Biological applications of electron microscope

1. Viewing of ultra structure of the cell organelles is made possible.
2. Structural details of large biomolecules such as proteins can be studied.
3. Minute details of the topographic architecture is possible, which is of systematic value as in the case of pollen grains.

Limitations of electron microscope

Even though electron microscope is a great advancement over the light microscopes in terms of higher magnification and better resolution, there are certain limitations and light microscopes have their own useful role to play in the microbiological laboratory. The limitations imposed by electron microscope are.–

1. Hydrated specimens are not to be used, hence live specimens cannot be observed
2. Dehydration even if carefully done will cause some distortion
3. Artifacts may appear
4. As electrons have a low penetration power extremely thin sections are necessary.

MICROMETRY

Micrometry or microscopic measurements refer to measurements of microbes (length, breadth), seen under the light microscope. Many a time, the size of the microbes is an essential criterion besides their morphological characters. The following are necessary to carry out microscopic measurements.

1. A good microscope with 10X eye piece and 10X, 40X and 100X objectives
2. Ocular micrometer
3. Stage micrometer

1. A good microscope (with straight tube) : is necessary to carry out the measurements.

The eye piece may be 10X or 15X with objectives of 10, 40 and 100. It should be noted that when measurements are to be carried out at each magnification, the original calibration (see later) conducted should be at the same magnification.

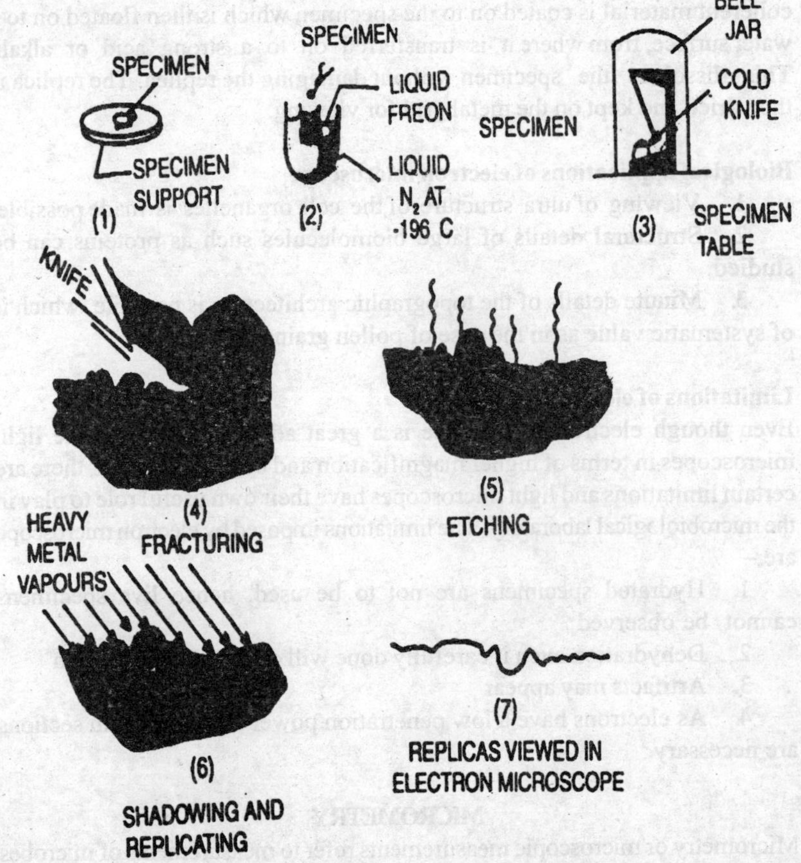

Fig. 3.18 Microscopy
Procedure for freeze etching replicas

2. **Ocular micrometer :** This is a small circular disc of glass which has graduations engraved on one surface. The distance between the lines is fixed, but the lines are arbitarily engraved in the sense it is not a standard measurement of mm, microns etc. In other words the lines on the ocular micrometer do not indicate any measurement of the length.

3. **Stage micrometer :** Also known as objective micrometer, the stage micrometer is a glass slide that has inscribed lines, which are exactly 0.01 mm (10 micrometer) apart. The lines can be easily seen when the slide is placed on the stage and viewed through the microscope.

Procedure for measurement

The first step in the measurement is the calibration of lines of ocular with that of stage micrometer.

What is calibration ? This refers to comparison or super imposition of scales of ocular as well as stage micrometer. This is necessary to determine how many graduations of ocular coincide with one graduation of the stage micrometer. It has already ben said that the lines of the ocular are arbitrary, while the lens of stage micrometer, are exactly 10 micrometers apart. So while using only the ocular micrometer for measurement, it becomes necessary to know as to what is the distance between graduations.

At this stage it can be said, that instead of the cumbersome procedure of calibration why not use only the stage micrometer which has the measurements ? This is difficult because the microbes then have to be mounted directly on the stage micrometer which is always not possible. Hence, calibration of ocular with the stage micrometer.

Procedure for calibration

The First step is the installation of ocular micrometer into the ocular lens. This can be done by unscrewing the top lens in the ocular, placing the micrometer inside the tube and screwing back the lens. In some of the latest microscopes, the ocular micrometer comes with a ring and it can be easily inserted into the bottom of the ocular tube.

The next step is placing the stage of the micrometer and focussing it properly. After focussing, the stage micrometer has to be adjusted so that its graduations align with the graduations of the ocular properly. After the graduations are aligned, we count how many ocular divisions equal one division (0.01 mm) of the stage micrometer. If (as shown in the figure) seven ocular divisions equal one division of stage micrometer, the value for each division of ocular is 0.01/7 or 0.0014μm. In other words, each ocular division is about 1.43 μm apart. Having known this, the stage micrometer may be replaced with the slide containing microbes for measurements. To determine the size of an organism, all that one has to do is to count the number of graduations, the microbe measures, and to multiply this number with 1.43 μm to give the actual size.

Note : Calibration has to be made for each pair of ocular and objective lenses, as the magnifications vary. Ideally, it is better to use the same microscope used for calibration, also used for measurements.

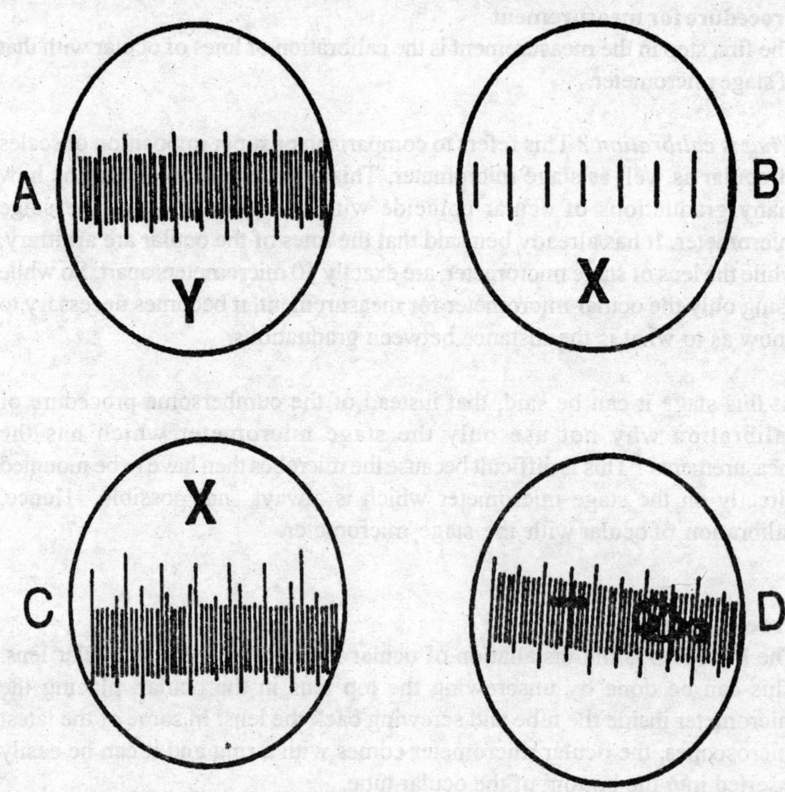

Fig. 3.19 Microscopy
Calibration of ocular micrometer

VIRUSES

The inhabitants of microcosm have a profound influence on human life. They are both friends as well as foes of man. While many have a prominent role to play in human welfare, others have caused immense harm to human life. Among the most important of microbial enemies of human beings, viruses occupy a prominent place. Simple and tiny enough to pass through the minutest bacterial filters, yet potent enough to change the destiny of human life, viruses are today the centre of attention in causing one of the deadliest of human diseases. What are these viruses? We will try to find answers to some of these in this chapter.

Discovery of Viruses

Adolf Meyer (1886), a dutch Agricultural Chemist, observed mottling disease in the leaves of tobacco plant and named it *Mosaikkranket* (Mosaic). He was able to demonstrate that the mosaic disease could spread from plant to plant if the juice of the infected leaves is applied to healthy leaves. He also showed that the juice even after passing through double filter paper retained the infectivity. He further demonstrated that heating the juice to 80°C would render it ineffective. He, therefore, concluded that the bacteria are responsible for the disease.

A further step in finding out the causative agent of tobacco mosaic disease was taken when lwanowsky (1892), a Russian scientist, showed that the juice even after passing through the fine bacterial filter (chamberland filter candles), retained the capacity of infection. This means that bacteria are not responsible for the disease. But when the filtered sap was put on the cultured medium no living organism grew. The filterable agent was called a Virus. The term virus was, however first introduced by Louis Pasteur to indicate the cause of canine rabies. But it was a general term used for various infectious agents.

Beijerinck (1898), a Dutch microbiologist, showed that the infective agent could diffuse into agar gel, like a liquid and he called it *Ucontagium vivumfluidum* (contagious fluid).

Loeffler and Frosch (1898), showed that infective agent of foot and mouth disease in cattle (like tobacco mosaic agent), passes through bacterial filter, and neither can it be seen under the microscope not can it be cultured on the culture media.

Viruses attacking bacteria were first discovered by a British scientist, Twort, and later in more detail independently by a French scientist 'd' Herelle, who named them bacteriophage. Subsequent studies showed the existence of hundreds of other viruses in plants, animals and human beings.

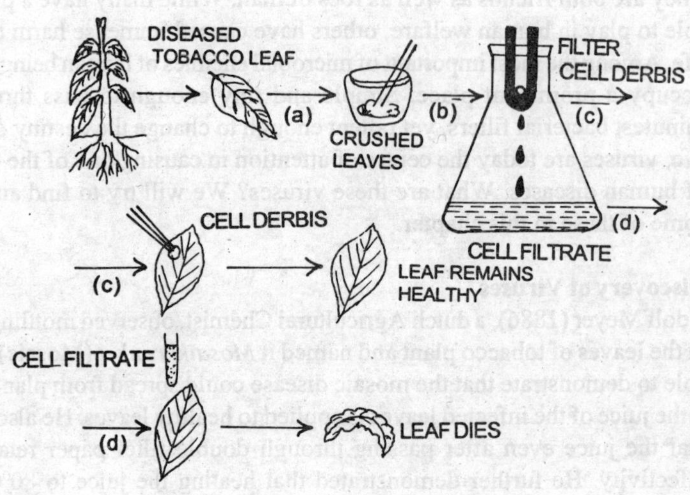

Fig. 4.1 Viruses

(a) Iwanowski's discovery of the filterable virus. Infected tobacco leaves were crushed and filtered through Chamberland's bacterial filter (b). The cell debris applied to healthy leaves without effect (c). When the clear filtreate was placed on the leaves (d). The leaves shriveled and died. This experiment showed the presence of a "filterable virus" that was smaller than any known bacteria

Schelsinger (1933), was the first to study the chemical composition of a virus. He showed that the bacteriophage consisted of only nucleic acids and proteins. A few years later, Stanley (1935), isolated tobacco mosaic virus in paracrystalline form. Later, Bawden and Pirie, (1938), identified the chemical nature of TMV and other viruses. Gierr and Schramm (1956), showed that the nucleic acid in viruses is the actual infective agent. Viroids and satellite viruses were discovered by Kassanis (1966), and Djcner and Raymer (1967), respectively.

Some important milestones in the history of the study of viruses (Virology) arc as follows:-

Year	Event	Name of the Scientist
1880	Discovery of rabies	Louis Pasteur
1886	Juice of tobacco mosaic is infectious	Adolf Mayer
1892	Infectious agent of Tobacco mosaic can pass through bacterial filter	Iwanowsky
1896	Concept of Contagium vivum fluidum proposed	Beijernick
1898	Foot and mouth disease in animals	F. Loeffler and A. Frosch

Year	Event	Name of the Scientist
1915	First viral infection of bacterial discovered	F.W. Twort
1917	The term 'bacteriophage' coined for the first time	d' Herelle
1935	Virus isolated in crystalline form for the first time	W.M. Stanley
1938	Chemical nature of TMV discovered	F.C. Bawden and N.W. Pirie.
1939	Viral mutations discovered	M. Delbruck
1946	Electron microscope study of plant virus	W.M. Stanley
1949	Cultivation of Polio virus in tissue culture medium	J. Enders
1951	Discovery of cyanophages	R.S. Shafferman and M.E. Morris
1957	First successful polio vaccination	J. Lindermann
1959	Ultrastructure of Ts bacteriophage	S.G. Brenner
1967	Discovery of Viroids	Dinner and Raymer
1973	Discovery of Viral cancer in primates	D. Schidolvski

1976	DNA Plant Viruses study	R.J. Sheperd
1979	Replication in vaccinia Virus	L.A. Gaurina and J.R. Kates

Are Viruses living non living?

Viruses being the simplest and most primitive, it is difficult to assign to them any rank in the living kingdom. It is also not easy to define the features of viruses within the accepted frame work of living beings. Hence, it is better to regard them as an intermediate stage between living and non living. Some of the characters of living as well as non living, exhibited by viruses are as follows:-

Characters of living beings

1. Viruses have genetic material (DNA or RNA).
2. They mutate.
3. They can grow.
4. They can be transmitted form one host to another.
5. They are capable of multiplication within a host.
6. They react to heat, radiation and chemicals.
7. They show irritability.
8. They bring about enzymatic changes in *vitro.*
9. They are able to infect and cause disease to living beings.
10. The DNA and proteins of viruses are similar in composition and structure to those of higher organisms.

Characters of non living

1. They can be crystallized like an ordinary chemical and stored in a bottle or test tube indefinitely.
2. Outside the host, viruses are inert.
3. There is no cell wall, membrane or cytoplasm.
4. There are no cell organelles, and there is no metabolism.
5. Sedimentation of viruses is according to their molecular weight like non living beings.
6. They do not have functional autonomy i.e., they are not capable of any function unless they obtain metabolic products from others.
7. Energy producing enzyme system is absent.

Some unique characters of viruses

The following characters are unique to viruses and are not seen either in living or non living.

1. Presence of only D.N.A. or R.N.A.

2. Capacity to reproduce from the sole nucleic acid.
3. They do not show cell division.
4. They use the metabolic machinery of the host cell to replicate.

In view of the above, as many virologists believe, it is better to regard that "viruses are chemicals in a test tube, but living beings inside the host"

Properties of Viruses

1. Viruses are called acellular as they do no have cellular organization like other microorganisms.
2. They are ultra microscopic (invisible under ordinary microscope).
3. The genetic material in viruses is either DNA or RNA but never both.
4. Viruses are obligate parasites. They can not be cultured on inanimate media.
5. Viruses can be cultured on cell media (bacteria, cells of chick embryo etc.,).
6. Viruses are filterable; they can pass through bacterial filters. They can, however, be filtered on molecular filters.
7. Even though viruses are made up of nucleoproteins, they lack the enzyme necessary for their synthesis. For this, they (Viruses) depend on the host enzymes.
8. Viruses do not show cell division.
9. They can be crystallized.
10. Viruses can be transmitted from one host to another, either directly or through vectors.
11. They can be transmitted from one host to another either directly or through fectors.
12. The capside (outer coat) of viruses is mostly made up of proteins except in some animal viruses, where polysaccharides are also present.

Classification of Viruses

A variety of criteria are used in classifying the viruses. Earlier, viruses were classified into several categories based on their hosts.

1. Plant Viruses – Infect plants
2. Animal Viruses – Ifect animals
3. Bacteriophages – Infect bacteria
4. Cyanophages – Infect cyanobacteria (blue green algae).
5. Mycoviruses – Infect fungi
6. Mycoplasma viruses – Infect mycoplasma
7. Phycoviruses – Infect algae.

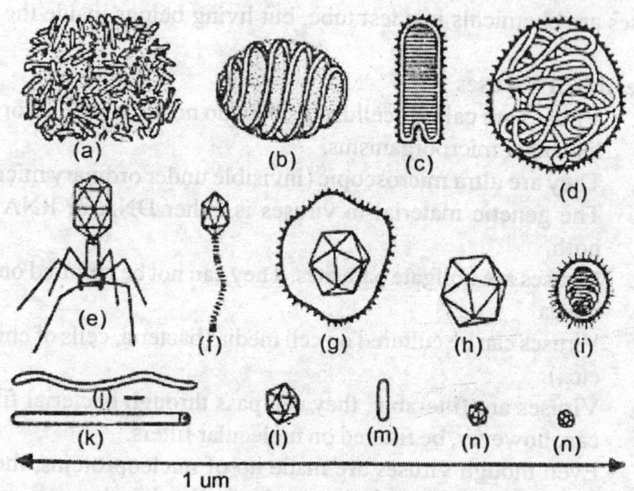

Fig. 4 .2 Viruses

Diagram of virus morphology and size range (a). Poxvirus (vaccinia) (b). Poxvirus (orf). (c). Rhabdo virus, (d). Paramyxovirus (Mumps virs), (e) T-even phage,

(f). Flexous taialed phase

Modern system of classification of viruses proposed by Lowaff (1966), and accepted by the provisional committee on nomenclature of viruses (PCNV), is based on the following criteria.

(i) Type of Nucleic acid – DNA *or* RNA.
(ii) Symmetry of the capsule – Helical, cubical and complex.
(iii) Presence or absence of envelope around the capsule.
(iv) Diameter of the helical capsule.
(v) Number of capsomeres (Units) in cubical viruses.

The following is an outline of the classification of viruses as proposed by Lowaff, Home and Toumier (1966).

This system is known as the LHT (Lowaff, Home and Toumier) system based on the name of the scientists who proposed the classification.

Phylum ***VIRA*** (includes all viruses) divided into two sub Phyla.

Sub phylum *Deoxyvira* (DNA viruses)

Sub phylum *Ribovira* (RNA viruses)

Sub phylum *Deoxyvira is* divided into three Classes–

Class 1. Deoxyhelica : Includes viruses with helical symmetry. It has only one order, Chitovirales (enveloped), and one family Poxviridae.

Class 2 *Deoxycubica* : Includes viruses with cubical symmetry. It has two orders *Haplovirales* (no envelope) and *Peplovirales* (enveloped). Haplovirales as five families – *Microviridae. Parvoviridae. Paploviridae, Adenoviridae* and *Iridoviridae.* Order peplovirales has only one family, *Herpesviridae.*

Class 3 *Deoxy binala* : Viruses have complex (Binal) symmetry. There is only one order *Urovirales* including a single family *Phagoviridae* (bacteriophages).

Sub phylum Ribovira is divided into two Classes Viz., *Ribohelica* and *Ribocubica.* Ribohelica includes viruses with helical symmetry, while ribocubica includes viruses with cuboid symmetry.

The class Ribohelica has two orders – Rhabdovirales andSagovirales. Rhabdovirales do not have an envelope and is further divided into two sub orders -- Rigidovirales (rigid viruses) with three families (Dolichoviridae, Protoviridae and pachyviridae), and Flexivirales (Flexible viruses), with three families (Leptoviridae, Mesoviridae and Adroviridae).

The order Sagovirales has enveloped viruses, and is divided into three families viz., Myxoviridae, Paramyxoviridae and Stomatoviridae.

The class Ribocubica has two orders Gymnovirales (viruses without envelop) with two families (Napoviridae and Reoviridae), and Togavirales with a single family Arboviridae.

PHYLUM VIRA

Subphylum Deoxy vira

Class	:	Deoxyhelica	
		Order :	Chetovirales
Class	:	Deoxycubica	
		Order :	Haplovirales
			Peplovirales

Class : Deoxybinala
Order : Urovirales

Subphylum Ribovira
Class : Ribohelica
Order : Rhabdovirales
Sagovirales
Class : Ribocubica
Order : Gymnovirales
Tugavirales

In another system of classification proposed by Casjens and King (1975), viruses are classified based on – type of nucleic acid, presence or absence of an envelop and site of assembly of virus particles in the host (whether in nucleus or cytoplasm). They have classified viruses into four divisions ssRNA viruses, dsDNA viruses and ssDNA viruses (For more details see Casjens and King J., 1975 - Annual Review of Biochemistry 44:555 - 611).

Structure and chemical composition
Viruses are not cellular, hence they do not have any cell wall, membrane and cytoplasm. They are extremely small and smaller than bacteria. The technical name for a virus particle is Virion. Virions vary in size. The smallest virus (causing foot and mouth disease) has a diameter of IOnm. Some of the large viruses may reach the size of small bacterium. For e.g., Pox virus is about 250nm. Lympho granuloma virus is 300 - 400nm in size.

Viruses vary in shape also. They may be rod shaped, bullet shaped, oval, irregular or pleomorphic. Based on shape, vinises may be categorized into two groups – polyhedral forms (Adeno virus), and helical forms (Tobacco mosaic virus).

Structurally each virion is very simple. It consists of a nucleic acid core surrounded by a protein coat (capsid), to form the nucleocapsid. The capsid is made up of many independent units called Capsomeres. The nucleocapsid may be naked or as in some cases surrounded by a loose membranous envelop. The capsid offers protection to the nucleic acids against the action of nucleases.

Shape of capsid The Capsid mainly exhibits three kinds of symmetry or shape. These are– polyhedral, helical and complex (binal). Polyhedral and helical viruses may or may not have envelopes.

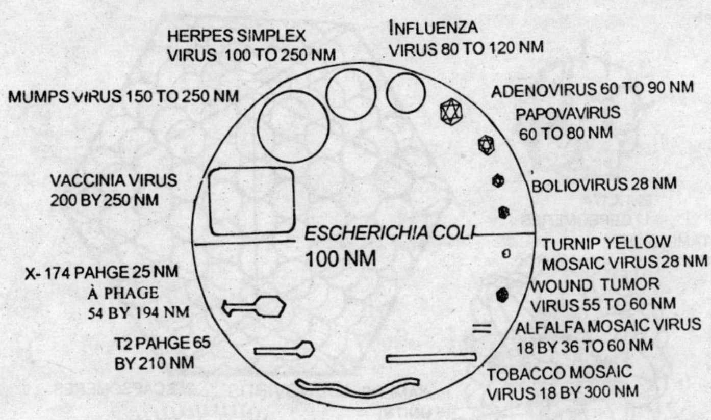

Fig. 4.3 Viruses

Diagrammatic comparison of size of viruses and related individuals. The largest circle, represents the diameter of E.coli, about 1.0µm. The other organisms are drawn approximately to the same scale

Polyhedral capsids are also called icosahedral and have tetrahedral, octa hedral or Polyhedral (more than eight) corners or vertices and 20 facets or sides. The shape of each facet is like an equilateral triangle. As has already been mentioned, each capsid is made up of a number of capsomeres and each capsomeres has a number of monomeres which form polygonal rings with a central space of up to 40A.

The capsomeres are of two types – pentameres and hexameres. The pentamere (pentagonal capsomere), is made up of five monomeres while hexamere has six monomeres. The monomeres are held together by bonds (e.g. Turnip yellow virus).

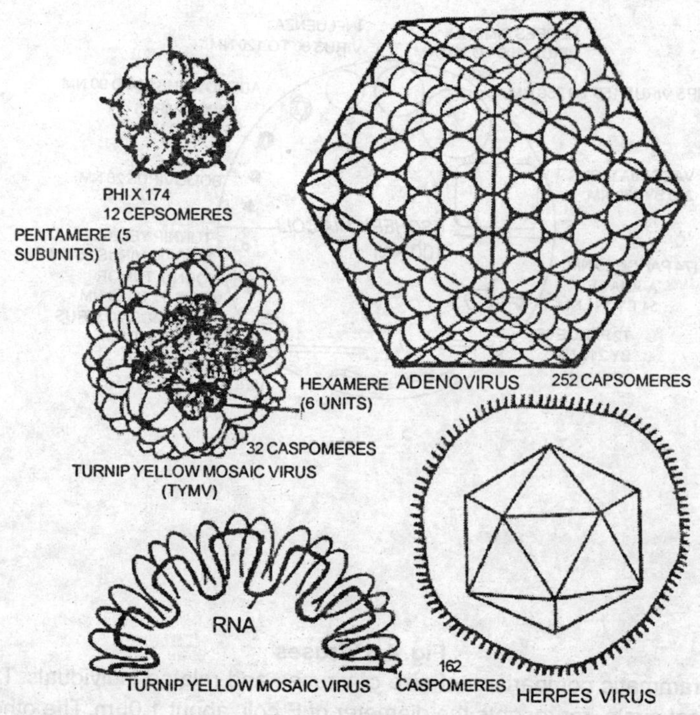

Fig. 4.4 Viruses
Polyhedral (icosahedral) viruses

The helical capsids have monomere arranged in the form of a helix around a single rotational axis. The monomeres curve into a helix because they are thicker at one end than the other e.g. TMV, Influenza virus etc.

The complex capsids are divided into two categories. Those without identifiable capsids and those with capsids to which additional structures are attached. In vaccinia virus, there is no identifiable capsid.

Bacteriophages of T even series have capsids with additional attachments. Usually the capsid has two regions – head and tail. The head capsid is a icosahedron and may be made up of 2000 identical subunits. The tail is connected to the head through a collar. At the free end, the tail has a base plate with 6 corners. At each corner is a spike. In addition, from each corner of the hexagonal plate a long tail fibre is given out.

Chemically the capsid is made up of proteins. Each capsomere in TMV is made up of 158 amino acids.

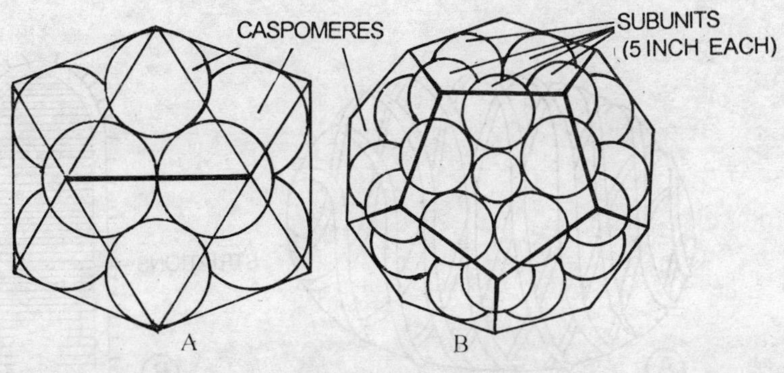

Fig. 4.5 Viruses

A cubic virus fX174 (a). A particle consists of 12 capsomeres arranged in icosahedral symmetry, (b). Six capsomeres, each made up of five subunits

Capsid helps the virus – in offering protection to the nucleic acid core, giving a stable shape and helping in the initial attachment of the virus to the host cell.

Structure of envelop

Some viruses have a membranous envelop external to the capsid. The envelop is about 100–150A^0 thick and is found in some icosachedral and helical animal viruses (also in some plant viruses and bacteriophages). The membrane is actually a portion of the host cell membrane which the .viruses acquire when they are released from the host cell after multiplication. The memberane has the typical bilayer structure with phospholipids and embedded proteins. In addition to the above, membranes may also have a significant amount of carbohydrates like galactose, mannose, glucosamine and galactosamine etc.

Nucleic acids

The core of the virion is made up nucleic acids. A virus has only DNA or RNA, never both together. Within these parameters, however, considerable diversity exists. The nucleic acids may be double stranded, single stranded, linear or circular. Some have plus polarity; other may have minus polarity. Four types of nucleic acids are found in viruses with reference to the number of strands. These are:

(i) Single stranded DNA'(ssDNA) E.g. Colliphage virus
(ii) Double stranded DNA'(dsDNA) E.g. Herpes virus
(iii) Single stranded RNA-(ssRNA) E.g. TMV
(iv) Double stranded RNA'(dsRNA) E.g. Reovirus

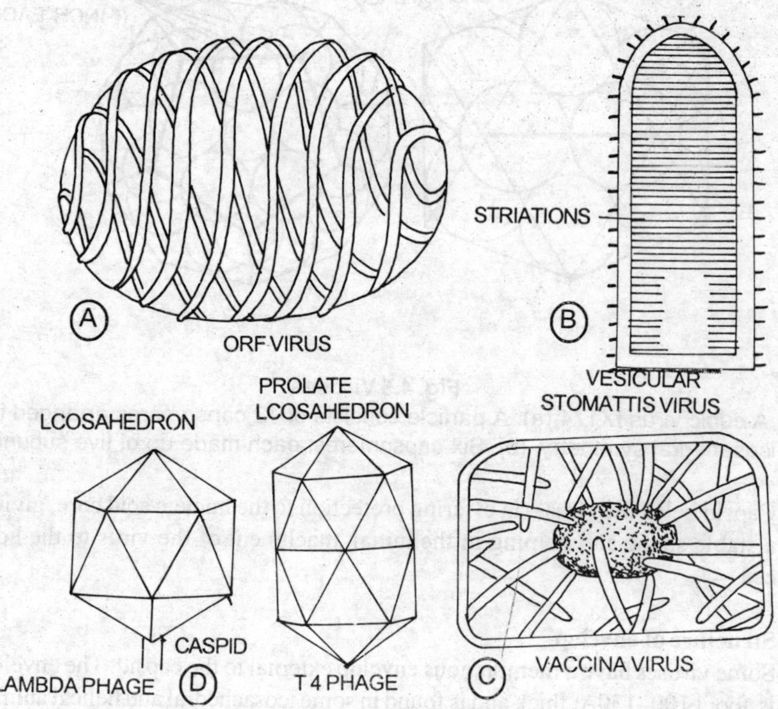

Fig. 4.6 Viruses
Complex viruses

Single stranded DNA is found in ϕX 174 virus. It was first discovered by Sinsheimer et al, in 1950's. The ssDNA may be linear (paroviruses), or circular (ϕX174 virus). The ssDNA becomes double stranded during replication and at that time it is known as the replicative form.

Double stranded DNA is found in a number of animal viruses and bacteriophages. It has a variety of forms – linear (bacteriophages), cross linked (vaccinia virus) or closed circular duplex as in papova viruses.

Single stranded RNA is found in a variety of animal viruses and icosahedral plant viruses. The strand may be plus (infectious), as in RNA bacteriophages, togavinises etc., or minus (non infectious), as in rhabdoviruses and paramyxoviruses. Plus ssRNA acts directly as mRNA and translates proteins on the ribosomes of bacteria, whereas minus ssRNA first transcribes an mRNA and then it (mRNA) translates proteins (hence called non infectious).

Double stranded RNA is found in animal viruses like reovirus, blue tongue virus etc. The ssRNA has 10 or more segments.

The nucleic acid core of viruses may be summarised as follows

(1) Plant viruses have only RNA (ss or ds) with the exception of cauliflower mosaic viruses (DNA virus).
(2) Animal viruses have RNA (ss or ds) and DNA, (only ds no ss).
(3) Bacteriophages have DNA (ss or ds). However, most phages are DNA viruses .

Replication of viruses

Like any other living being, viruses also multiply but with a difference. They cannot multiply on their own. They do not reproduce by division independently because the viral nucleic acid cannot duplicate by itself. There are no raw materials to produce fresh nucleotides, neither is there an enzymatic machinery to assemble them together into a duplicate molecule.

Viruses multiply only when they are in a living cell. The DNA or RNA of the virus takes control of the host cell metabolic machinery and new viral particles are produced utilizing the raw materials from the host cell.

Multiplication of tobacco mosaic virus (TMV), takes place inside the leaf cells of the host. The virus reaches the inside of the cell either through a wound or abrasion. The protein coat of the virus is dissolved by host enzymes and the synthesis of mRNA, proteins and genome as well as the maturation of the viral particles occur in the nucleus of the host cell and not in cytoplasm. The cells do not breakdown (no lysis) and the virus particles move into other cells of the plant through plasmodesmata. Viral particles are transmitted through sop from injured tisssues.

Replication of viruses is studied in great detail in bacteriophages. Two kinds of replication cycles occur in phages. These are the *virulent* or *lytic cycle,* and the *temperate* or *lysogenic cycle.* In the lytic cycle, the viral particles multiply inside the bacterial cell. After the formation of new viral particles the host cell breaks down (lysis), and the viral particles are released. In the lysogenic cycle, the viral genome (after entry) gets integrated with the bacterial genome and is called prophage. The prophage, multiplies along with the bacterial genome (no independent multiplication) without causing any damage to the host cell. At some stage, however, depending on environmental conditions, the prophage may separate from the bacterial genome and start a lytic cycle.

As the lytic bacteriophages (which efect E. coil) have been studied in great detail with reference to viral multiplication. The following account is mainly based on them. However, in other viruses also multiplication is same as in T even phages.

The main stages involved in the replication of T even phages are–

(i) Attachment of phage particle to the host.
(ii) Adsorption of virus particle.
(iii) Separation of nucleic acid from coat.
(iv) Penetration into the host.
(v) Effect of phage attachment on the host cell metabolism.
(vi) Replication of viral of genome (nucleic acid).
(vii) Synthesis of viral protein (capsid).
(viii) Assembly of new viral particles and
(ix) Release of viral particles from the host cell.

Attachment (of phage) to the host cell

Viral particles come into contact with host cell surface. The tail plate of the virus attaches to the surface of the host cell, and anchors itself firmly with the help of the tail fibres. The attachment of tail plate (of the virus) to the host cell, is a highly specific chemoreceptor regulated process. Certain protein receptors present in the tail capsid of the virus can recognise receptor sites on the host cell surface. It has been noticed that in *E.idwrichia coU,* the outer lipoprotein layer in the peptidoglycan covering has many receptors for pliage attachment.

Absorption of pliage

Initially the attachment of the phage to the host cell is reversible, but later it becomes irreversible (virus can not be removed). Some cations are known to play a role in the absorption of the Phage. The receptors of the host cell surface are complex polysaccharides which can bind the phage on antigenic specificity. The absorbed phage has its head perpendicular to the cell surface. If the host cells ar capsulated, and enzyme at the tail top of the phage hydrolyses the polysaccharides of the bacterial capsule boring a tunnel through which cell the wall can be reached. The receptor in phage is a mannose transport protein. The receptor in bacteria some times may be present in pili (as in males), in which case some coliphages are male specific i.e., they attach to the pili sex only.

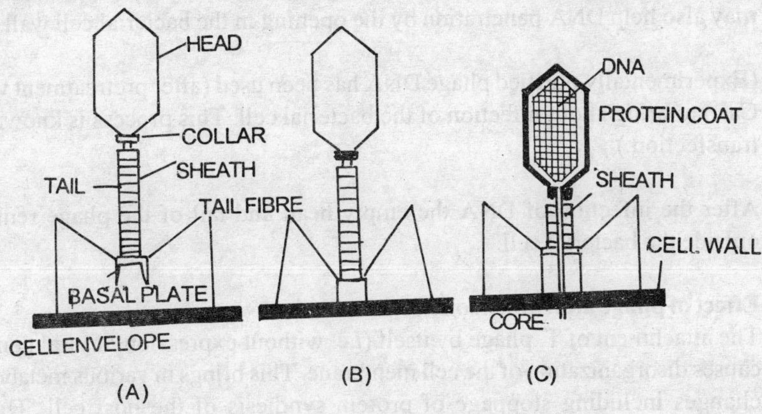

Fig. 4.7 Viruses

Mechanism of attachment and introduction of DNA intohost cell in a T-even bacteriophage. (a). The tail fibers get fixed to the surface of host cell. (b). The base plate is brought into contact with the cell surface. (c). The sheath contracts driving the tubular core through the cell wall of bacteria and the DNA from the head enters the host cell through the core

Separation of nucleic acid from coat

Hershey and Chase (1952), have shown that phage DNA carries the genetic information into the cell. Injection of DNA into the cell does not require any energy from the host. After the DNA is injected, the coat remains outside.

The naked DNA temporarily loses the infectivity until the production of new viral particles. This temporary phase is called *eclipse*.

Penetration into the host

The nucleic acid together with some internal proteins is released from the capsid after the virus is irreversibly attached to the host cell. It penetrates into the cell at the sites were the inner and outer membrances are in contact with each and remains associated with the membrane physically. The phage DNA is protected against the membrane nucleases of the host by its asociated proteins and by DNA modifications. A highly specialilzed mechanism of injection of DNA is seen in the T–even phages. Electron microscopic studies have revealed that after the tail plate is fixed to the cell, the tail sheath protein becomes contractile and pulls the collar and the phage head towards the basal plate and pushes the tube through the cell wall. When the tube reaches the plasma membrane, DNA is ejected. The energy necessary for contracion is derived from ATP present on the phage tail. Lysozyme present in the tail may also help DNA penetration by the opening in the bacterial cell wall.

(Experimentally purified phage DNA has been used (after pretreatment with Ca^{2+}) to bring about infection of the bacterial cell. This process is known as transfection.)

After the infection of DNA the empty head and tail of the phage remain outside the bacterial cell.

Effect of phage attachment on host metabolism

The attachment of T. phage by itself (i.e. without expression of viral genes) causes disorganization of the cell membrane. This brings in various metabolic changes including stoppage of protein synthesis of the host cell. These changes sometimes lead to cell lysis even without viral multiplication.

Replication of viral nucleic acid

Viral nucleic acid molecules produce many replicas using the enzyme machinery of the host cell. Host cell nucleus is the site for the multiplication of viral DNA.

Protein Synthesis of virus

In order to produce the'capsid', fresh viral proteins are to be produced. First, some enzymes are synthesized. These produce proteins peculiar to phage called'early proteins' and later the late proteins' are produced which develop into subunits of head and tail. At this stage, bacterial metabolism including DNA, RNA and protein synthesis comes to a halt.

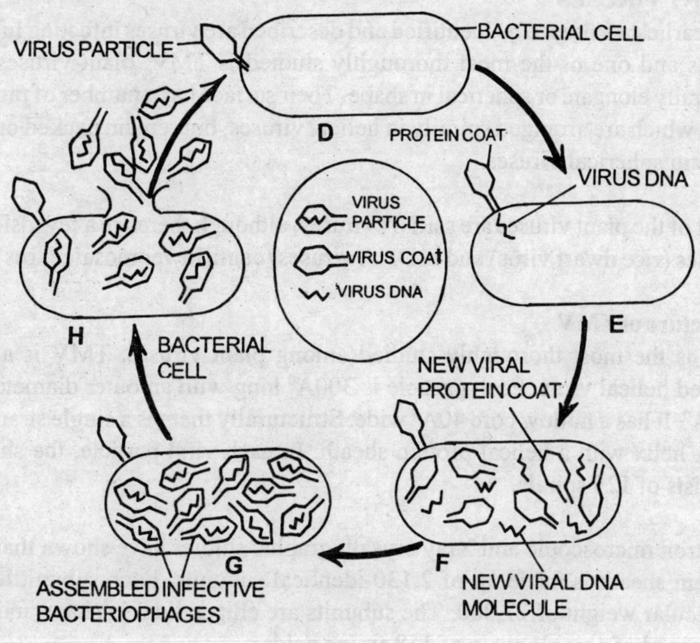

Fig. 4.8 Viruses
Attachment, adsorption and penetration (A-C) of a bacterial host cell by a T-phage; D-H stages of the cycle of infection of bacteria cells by bacteriophages

Assembly of new viral particles

Phage DNA and capsid proteins are synthesized separately in the bacterial celi. The phage DNA is then condensed and pushed into the head, and tail is added. The process of assembling the parts of a virus into a new virion is called 'maturation'.

Release of (new) viral particles

When the phage policies are being produced in the cell, the wall of the bacteria gets weakened considerably. Further, viral enzymes (produced in the bactrial cell) act on the cell wall resulting in its bursting open. As a result, phage particles are released outside and they infect other cells repealing the cycle.

PLANT VIRUSES

The earliest viruses to be identified and described are viruses infecting higher plants and one of the most thoroughly studied is TMV. plant viruses are generally elongate or spherical in shape. Their surface has a number of protein units which are arranged spirally in helical viruses, but remain packed on the sides in spherical viruses.

Most of the plant viruses are ssRNA viruses, although there are a few dsRNA viruses (rice dwarf virus) and dsDNA viruses (cauliflower mosaic virus).

Structure of TMV

This is the most thoroughly studied among plant viruses. TMV is a rod shaped helical virus. Each particle is 300A^0 long with an outer diameter of 180A^0. It has a hollow core 40A^0 wide. Structurally there is a single stranded RNA helix with a helical protein sheath. In each viral particle, the sheath consists of 129 spirals.

Electron microscopic and Xray crysallographic studies have shown that the protein sheath is made up of 2,130 identical subunits. Each subunit has a molecular weight of 17,000. The subunits are ellipsodidal in shape and are composed of a single chain of 158 amino acids.

The nucleic acid, which is in the centre is a single stranded RNA wilh a molecular weight of 2.4 million. The helix of RNA has a pitch of 23A^0 and a radius of 40A^0. There arc three nucleotides per particle of TMV.

SOME MAJOR GROUPS OF PLANT VIRUSES

Virus group	Nature of N.A.	Representative viruses	Properties of various
Tobacco mosaic	RNA	TMV; tomato aucuba Mosaic; Holmes rib grass; cucumber viruses 3 and 4	Rigid rods, 3000^{0}A long
Potato X	RNA	Strains of H, G, L etc,	Flexible rods,

		potato mottle; potato ring spot	about 5000A^0
Potato Y	RNA	Leaf drop streak; rugose mosaic; potato virus C	Flexible rods, about 7000A^0 long
Tobacco rattle	RNA		Rigid rods, 1750A^0 long
Tobacco etch	RNA		Flexible rods
Soyabean mosaic	RNA		Flexible rods
Tobacco necrosis	RNA		- do -, 170 A^0
Tobacco bushy stunt			- do -, 280 A^0
Southern bean mosaic	RNA		- do -
Turnip yellow mosaic	RNA		- do -
Potato yellow dwarf	RNA		Enveloped, bacilliform 750 X 3800A^0 diameter
Cauliflower mosaic	DNA		

Viral diseases of plants

Viral plant diseases may be systemic, or local. In systemic infections, the disease spreads all over the plant through the vasculture, while in local, the disease is restricted to certain areas. The following are some of the important symptoms of viral plant diseases.

1. **Mosaic :** Mixed light yellow and green pattern appear on the leaves making a mosaic pattern. This is a systemic infection, e.g., TMV, mosaic of cucurbits, mosaic of potato etc.

2. **Yellow :** Uniform chlorosis all over the leaf. It is also called chlorosis disease,e.g., Rice yellow.

3. **Enantions :** Small hair like outgrowths appear on the leaves called enantions. These outgrowths often accompany mosaic symptoms , e.g.. Bean enaiition mosaic.

4. **Vein clearing, vein banding and vein thickening :** Veins and veinlets become light yellow and transparent while the lamina remains normal. It is *vice versa* in vein banding–the whole lamina turns yellow, e.g., vein clearing in Lady's finger.

5. **Leaf curling or leaf rolling.** The infected leaves curl or roll back, e.g., leaf curl of papaya, tomato, etc.

6. Little leaf : In this, the affected leaves become small and crowded giving a rosette appearance, e.g., Little leaf of brinjal.

7. Stunting : The infected plants have overretardation of growth, e.g.. Bunchy lop of banana.

8. Breaking and greening of plants; Petals of flowers change colour, become small and green and appear variegated. Some times it may enhance the beauty of the flowers, e.g., Broken tulip flowers.

9 TiimiHirs : Some viral infections produce tumours on leaves or roots.

10. Witche's broom : Leaves get reduced and so also the internodes, resulting in crowded clusters looking like a broom.

Transmission of plant viruses

Viruses are transmitted from plant to plant in a number of ways. The chief methods of transmission of plant viruses are:

(1) Transmission by vegetative propagation

It is one of the chief methods of transmission, whenever plants are propagated vegetatively by means of cuttings, tubers, bulbs, rhizomes etc., viruses present in the other plant are easily carried over to the propagated plant. This method of transmission is chiefly seen in Potato, Strawberry, Raspberries etc.

Vectors and the Plant Viruses they Carry

Vector	Virus transmitted
Nematode	
Xiphinema index	Grape 'fan-leaf' virus
Fungi	
Olipidium brassicae	Tobacco mosaic virus (TMV)
	Tobacco stunt virust (TSV)
	Lettuce big-vein disease virus
Synchytrium endobioticum	Potato virus X (PVX)
Insects	
Myzus persicae (aphid)	Potato leaf-roll virus.
Nephotittix apicalis	Rice dwarf disease virus
Circudulina sp (leaf-hooper)	Maize streak virus.
Thysanoptera (thirp)	Wheat spotted - wilt virus.

Bemisa tabaci (white flies)	Tomato leaf-curl virus infecting tobacco

(2) Seed transmission

Transmission of viruses through seeds is not very common. It has been reported in very few cases such as: legumes, tomatoes, bean-mosaic, cucumber mosaic etc. They seem to come from ovules of infected plants. The viruses however, do not enter the embryo.

(3) Soil transmission

Soil transmission has been observed in few cases such as tobacco mosaic virus, potato mosaic virus, wheat mosaic virus, oat mosaic virus etc. These viruses live in the soil along with the plant debris after the harvest of the crop when sown in the same field.

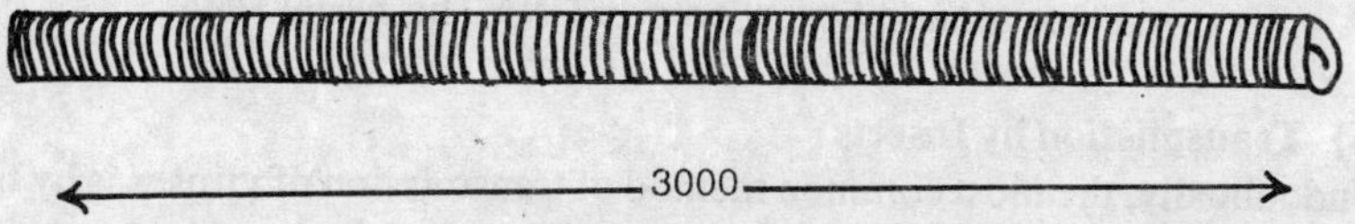

Fig. 4.9 Viruses

A rod of Tobacco mosaic virus, showing arrangement of spirals forming the protein sheath

(4) Transmission by contact

This is brought about in nature, by contact of one plant with the other. This is usually brought about by strong wind, which makes the infected part to come in contact with a healthy plant.

(5) Tobacco mosaic virus can remain infective in dry tobacco leaves as long as 25-30 years. They can be easily transmitted by the fingers of the smokers and also by the unburnt pieces of cigarettes, bi8dies, cigars etc.

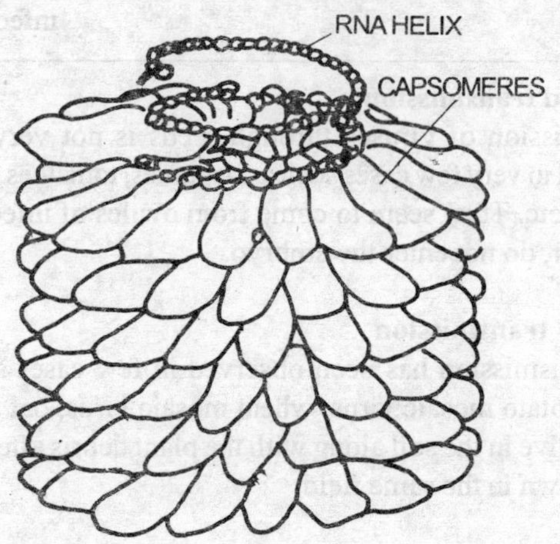

Fig. 4.10 Viruses

A helical virus, TMV. Proteinaceous capsomeres arranged in helical manner around hollow core of RNA The hilical RNA

(6) Transmission hy Insects

Undoubtedly, the most common method of transmission of viruses is by the activity of insect, which are termed vectors. Some of the insects which play an important role in the transmission of viruses are Scale Insects, *White* flies, Leafhoppers, Aphids etc. Aphids transmit more plant viruses than any other insects. Insects with their well developed mouth part carry viruses on their styles (Style borne virus), or may accumulate them internally, and after passing through the tissues, they are introduced again through the mouth part (circulative virus). Some of these viruses may multiply forming the propagative viruses. In some cases, the vector fails to infect a healthy plant soon after it is fed upon a diseased plant, but it can infect the healthy only after some time. This period of development of infectivity for the virus within the vector is called the *Incubation* period. This period varies with different viruses from hours to days.

There seems to be some relationship between the plant viruses and the insect vectors which transmit them. For eg:-Stunt virus of paddy is transmitted only by a Icaft hoper *Nepholetlix apicalii*.. Another type of leafhopper namely, *Circulifer tenellus* transmits only the curly top virus of beet root. The exact nature of the relationship between the virus and the vector is still unknown.

(6) Fungal transmission

Certain fungi also take in the transmission of viruses. For eg:- the root infecting fungus *Olipidiwn hrassicae* ,transmits at least 2 plants viruses namely, (1) Tobacco necrosis virus and (2) Tobacco stunt virus.

Control

1. Field sanitation measures
2. Growing disease resistant varieties.
3. Diseased plants should be uprooted and burnt to avoid further infection.
4. Insect vectors should be kept in check by spraying.

ANIMAL VIRUSES (including human)

Animal viruses bring about a number of diseases in animals and also in human beings. Animal viruses have either DNA or RNA as the genetic material. DNA is mostly double stranded while RNA is either single standed or double standed. Given below is an account of some animal viruses.

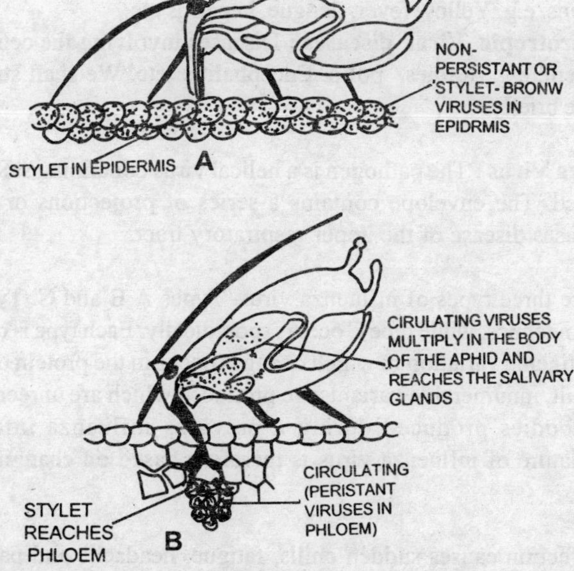

Fig. 4.11 Viruses

Diagram showing acquisition of viruses by aphids, **A.** Stylet borne virus, **B.** Circulating virus

Foot and mouth disease virus

This is a pathogen of cattle. .The virus is a single standed RNA particle having about 7,500 bases with a coding capacity of 2,500 amino acids. *The* capsid is icosahedral.

Rabies Virus

The pathogen is called a rhabdovirus and is bullet shaped. It is found in plants, invertebrates and vertebrates. In dogs it causes rabies, the virus is enveloped which is made up of two layers of lipids. The hereditary material is ss RNA. (for more details see under human viruses) Another example of habdovirus is the vesicular stomatitis virus which is a mild pathogen of cattle.

Human pathogenic viruses

These are artificially classified into the following four groups based on the organs in which disease symptoms are produced.

1. **Pneutropic Viral diseases :** Diseases involve respiratory tracts, e.g. Influenza, Common cold etc.
2. **Dermotrophic viral diseases :** Diseases involve mainly skin e.g. Measles, Chicken pox. Small pox, mumps etc.,
3. **Viscerotropic viral diseases :** Diseases involve blood and visceral organs, e.g. Yellow fever, dengue, hepatitis etc.,
4. **Neurotropic Viral diseases :** Diseases involving the central nervous system, e.g. Rabies, polio. Encephalitis etc. We shall study some of these briefly.

Influenza Virus : The pathogen is a helical virus containing ssRNA and it is enveloped. The envelope contains a series of projections or spikes. The virus causes disease of the upper respiratory tract.

There are three types of influenza virus–Types A,B and C. Type A and B causes epidemics, while type C occurs sporadically. Each type is characterized by its antigenic variation in which changes occur in the protein of the capsid. As a result, innumerable variants are produced which are unrecognizable by the antibodies produced during a previous influenza infection. The nomenclature of influenza virus is therefore based on changing antigenic pattern.

Viral infection causes sudden chills, fatigue, headache and pain. Fever of upto to 103^0 to 104^0F is very common, influenza A is treated with amantadine, a synthetic drug which is known to prevent viral penetration into the host cells.

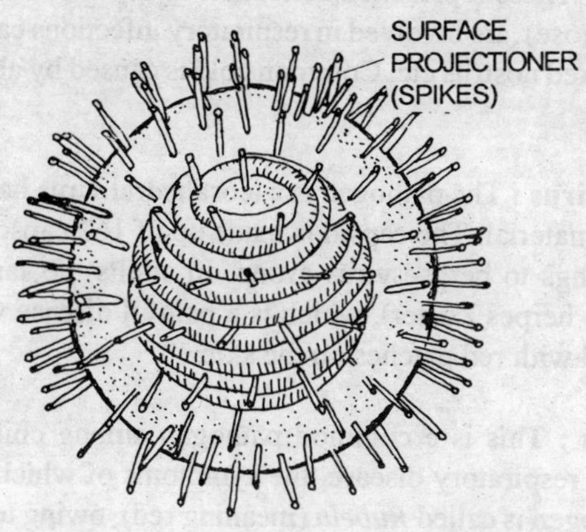

Fig. 4.12 Viruses
A typical myxovirus (influenza virus)

Adenoviruses : These are a group of about 35 icoscahedral particles having ds DNA. The viral particle is about 60-90nm indiameter having a dense central core and the capsid composed of 252 capsomeres (240 hexons and 12 pentons).

The viruses owe their name to the adenoid cells from which they were first isolated. They multiply in the nuclei of the host cells.

Adenoviruses cause diseases in respiratory tracts, eyes and meninges. Type 8 adenovirus is the main agent of Kerato conjunctivitis inflammation of the eye.

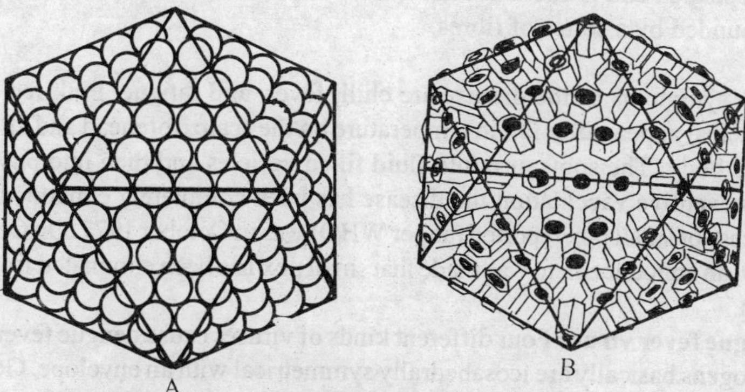

Fig. 4.13 Viruses
(a) Adeno virus (b) Herpes virus

Rhinoviruses : These represent a collection of about 100 icosahedral RNA viruses (Rhinonose) ,are involved in respiratory infections causing headache, sneezings blocked nostrils etc. Common cold is caused by about 15-29 types of rhinoviruses.

Chicken pox virus : The pathogen is a icosahedral virus having ds DNA as the hereditary material. The capsid is made up of 162 capsomeres. Chicken pox virus belongs to herpse virus group. In adults the same virus causes herpes (Virus - herpes Zoster), which is a painful disease where the spinal cord is affected with red patches on the skin.

Measles Virus : This is a common pathogen among children, causing a communicable respiratory disease, the symptoms of which develop on the skin. The pathogen is called *Rubela* (meaning red), owing to the appearance of red rashes on the skin.
Rubela is a helical ssRNA virus with an envelop in addition to the capsid. An effective vaccine is available against the disease.

Mumps virus : This is disease in children mainly affecting the parotid glands (hence the technical *name-epidemic parotitis* for the disease) There is a swelling visible on either side of the cheek. The virus has ss RNA and is helical having an envelop in addition lo 'capsid'. *This is similar to measles virus except that there are no nenraminidase spikes on the envelop.* A vaccine is available for the disease.

Small pox virus : Commonly called variola virus it has been eradicated the world over. The virus causing the dreadful disease has a complex archiecture, brick shaped and is about 270nm in size. It has no envelop but the capsid is surrounded by a series of fibres.

Early symptoms of the disease are chill, fever, and fatigue. Pink red spots (mancles) appear after fall in temperature on the scalp, forehead and all parts of the body. The spots turn into fluid filled vesicles and then into pustules. By preventive vaccination the disease has been completely contained. The last case of small pox reported (as per WHO) was in October 1977. On October 1979, an announcement was made that small pox has been eliminated from the earth.

Dengue fever virus : Four different kinds of viruses cause dengue fever. The pathogens basically are icosahedrally symmetrical with an envelope. Genetic material is RNA. Viruses are transmitted to human beings through mosquito bite. Dengue fever is a debilitating disease accompanied by severe pain in the limbs.

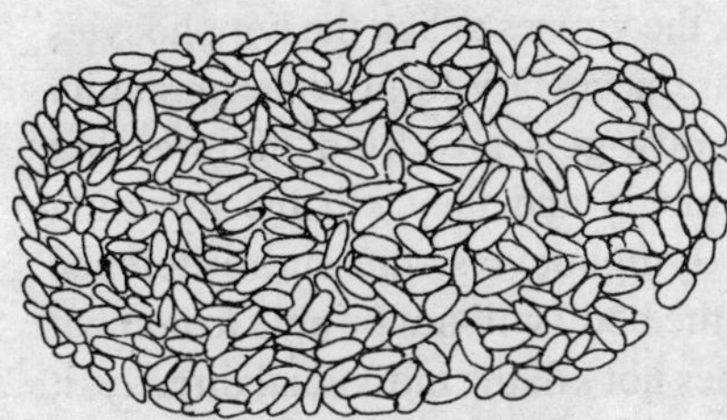

Fig. 4.14 Viruses
Poxvirus a brick-shaped virus

Rabies virus : The rabies virus not only causes disease to dogs, but to human beings also with a high rate of mortality if not treated properly. Besides dogs, rabies virus also occurs in cats, horses, rats etc.

The virus has RNA as the genetic material and is enveloped. It is rounded at one end and flattened at the other to give the appearance of a bullet. The virus gains entry into the human tissue through wound or abrasion caused from an infected animal. Though the disease is mainly transmitted through saliva, urine, lymph, or even milk may be equally infective. The virus after gaining entry spreads rapidly through the neural pathway. Disease symptoms include, aggressive behaviour, difficulty in swallowing, excessive salivation, dislike for water (hydrophobia) etc.

Poliovirus

One of the smallest among viruses, the polio virus (diameter 27-30nm) has an icosahedral capsid with a ssRNA as genetic material. The multiplication of the virus is phenomenal; within a six hour cycle it produces as many as 10,000 new particles.

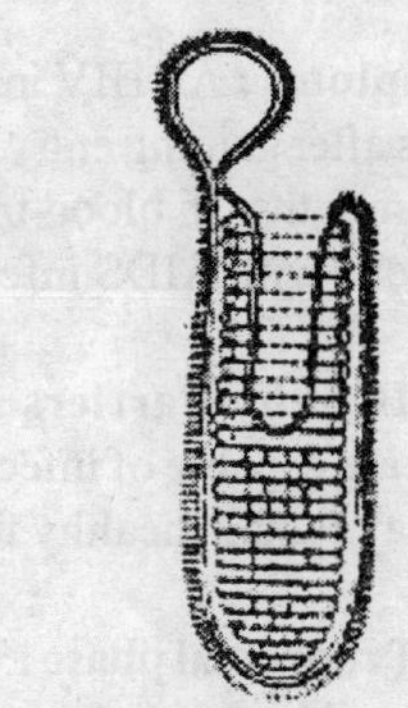

Fig. 4.15 Viruses
Rhabdovirus, a bullet-shaped virus

The virus after gaining entry through contaminated food or water multiplies in the tonsils and the lymphoid tissues. After entering the blood stream, the virus localizes in the grey matter of the spinal cord. (Greek polis = grey). Three types of polio are known to exist–Brunhidle strain-epidemics and paralysis; lansing strain occurs sporadically but causes greatest number of paralytic cases and Icon strain occurs occassionally confining itself to intestine rarely causing paralysis. A very efficient vaccine was prepared by Jonas Salk and his associates by using formaldehyde inactivated polio viruses. Another vaccine which can be taken orally is the one prepared by Albert Sabin which consists of attenuated viruses. This is known to produce more antibodies. Polio vaccine is routinely given to children as a part of general immunization.

AIDS

AIDS or Acquired Immunity Deficiency Syndrome is caused by a virus belonging to Retrovirus group. It has been given the name HIV (Human immunodeficiency virus). In this disease, the viruses attack the lymphocytes (T4), and destroy them. Lymphocytes are necessary to provide immunity against any disease. Whenever our body gets any kind of infection lymphocytes are produced in large numbers and they kill the infecting microbes. With the destruction of the lymphocytes, the entire defence mechanism (immunity) is disturbed and the body will not be able to combat even ordinary infections. While AIDS does not kill the individual directly, it renders the body weak and susceptible to all kinds of infections.

The first few AIDS cases were reported in 1981 in Atlanta, USA, since then it has spread to all countries including India.

Transmission of AIDS : The following are the methods of viral transmission, (a) through sexual intercourse with infected persons (Homosexuals are more succeptible), (b) through contaminated injecting needles, (c) through blood transfusion and (d) infants of infected mothers also get the disease through maternal blood.

Symptoms : All HIV infected persons need not show visible symptoms. The virus after gaining entry may remain dormant for a period ranging between 9-300 months. In blood transfusion cases, this may be between 4-14 months. Persons with AIDS infection can be classified into 3 categories.

1. **Healthy Carriers :** This is the first stage; there are no symptoms. But they are capable of infecting others. They may develop full blown symptoms or may remain healthy for a long period.

2. **Prodromal phase :** This is called AIDS related complex (ARC). This group has a milder from of the disease with fever, diarrhoea, weight loss and enlarged lymph glands. They exhibit depression of immune system.

3. **The End Stage :** Here AIDS is fully developed and the patient dies of incontrollable infections of lung, exertion, chest pain, fever, loss of weight, severe headache, loss of memory, convulsions, and development of sidii malignancy (Kaposis sarcoma).

Identifications of AIDS : Serological test of the blood by ELISA or western blot will indicate the presence of antibodies against HIV in infected patients.

Cure : Till now no cure has been found for AIDS. Drugs Azidothymidine,

Suramin etc. may give, a temporary relief. Attempts are being made to develop a vaccine against the virus. However, chances of contacting AIDS.may be avoided by taking the following precautions.

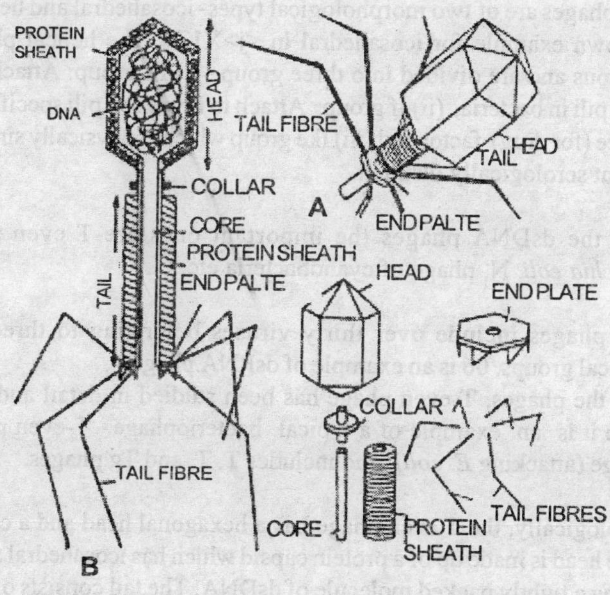

Fig. 4.16 Viruses. Morphology of T-even bacteriophage virus
A. External morphology; **B.** Diagrammatic representation of L.S. of a bacteriophage;
C. Various components of a bacteriophage

(a) Sterilizing the needles properly before injection or use of disposable syringes.

(b) Taken utmost care during blood transfusion; only HIV screened blood should be used. This is a moral step. Avoid sexual contact with several partners. Sexual contact to be had with only one known partner. Homosexuality should be strictly avoided.

BACTERIOPHAGES (BACTERIAL VIRUSES)

Viruses that infect bacteria are called bacteriophages. The word phage means eating, hence bacteriophages are bacteria eaters. Bacteriophages may have ssDNA, dsDNA, ssRNA or dsRNA as the genetic material. The word bacteriophage was first coined by d'Herelle (1917).

ssDNA phages are of two morphological types–icosahedral and helical. The best known example for icosahedral in <j>X174. The helical phages are filamentous and are divided into three group– (i) Ff group: Attach to the F type sex pili in bacteria, (ii) If group: Attach to I type sex pili specified by the resistence (for drug) factor and (iii) Ike group which is physically similar to Ff group but serologically distinct.

Among the dsDNA phages the important ones are-T even phage of *Escherichia coli.* N_1 phage of cyanobacteria etc.

ssRNA phages include over thirty viruses belonging to three or four serological groups. 06 is an example of dsRNA phage.
Among the phages, T-even phage has been studied in detail and we shall describe it is an example of a typical bacteriophage. T-even phage is a coliphage (attacking *E. coli),* and includes T, T, and Tg phages.

Morphologically, the T even phage has a hexagonal head and a cylindrical tail. The head is made up of a protein capsid which has icosahedral symmetry enclosing a tightly packed molecule of dsDNA. The tail consists of a hollow core surrounded by a contractile sheath. At the free end of the tail is present an end plate with a number of tail fibres. The end plate also has a number of spikes. The tail connects to the head with help of a disc or collar. The size of the heades 95nm X 65nm, while the tail is 115nm long with a diameter of 15nm.

The genetic material is a DNA with a molecular weight of 120 x 10* daltons. Some virologists believe that the DNA has hydroxymethyl cytosine instead of the normal cytosine.
(Life cycle and multiplication of bacteriophages is discussed under virus multiplication.)

CYANOPHAGES

Viruses that attack cyanobacteria(blue green algae) are called cyanophages. The first suggestion that viruses may be the cause of sudden inexplicable disapperance of blue green algal blooms was first made by Krauss (1961). Earlier (1960), Lewin reported a phage attacking *Spirochaeta rosea* (considered to be a blue green alga by some phycologists).

sThe first cyanophage was isolated in 1963, by Saffermann and Morris . Since then there have been several reports of cyanophages including one from India (Singh and Singh 1967).

NOMENCLATURE

Cyanophages are named after the hosts they attack. If they attack more than one host, the names of the hosts are abbreviated. For example, cyanophage attacking *Lyngbya, Phormidium and Plectonema* is named LPP-I taking the first letters of the three hosts.C-l cyanophage attacks *Cylindrospermum. So* far about 14 genera of cyanobacteria are known to harbour viruses-These (hosts) are *Anacystis, Chroococcus, Microcystis, Nostoc, Lyngbya, Phormidiutn, Cylindrospermum etc.*

Structure

Cyanophages may be icosahedral or helical. The capsid protein has a molecular weight of about 4X *W* daltons. The LPP-virus has a distinct head and a short tail similar to a bacteriophages. The capsid head is about 600nm and the tail 200nm long. The capsid is covered by an envelop. The genetic material is DNA. It has a *C/G* ratio of 55-57%.

The infection and multiplication of cyanophages is almost similar to bacteriophages. The viral DNA multiplies in the host cell in between photosynthetic lamellae, and in the region where host DNA lies (nucleoplasm). It has been reported that LPP- 1 phage breaks down half of the host DNA, and incorporates it into its own DNA.

Viruses attack the host resulting in the rates of photosynthesis and respiration coming down. Lysis of the cells of the host is restricted to vegetative cells while heterocysts and spores are affected very little.

Isolation

Samples of sewage water are added to algal culture which is incubated for about a week and centrifuged. The supernatant is treated with chloroform and screened for phage activity by allowing it to grow on a test blue green alga.

Uses

Cyanophages can be used to destroy blue green algal blooms. Goryush and Chaplinskaya (1968), have reported the destruction of blooms by using cyanophages. Daft and Stewart (1971), have also demonstrated a similar phenomenon

ANIMAL VIRUSES

Family	Capsid symmetry and Assembly site	Viron diameter (nm)	Capsomera number	Nucleic acid	Envelop	Typical agents	Disease caused
1. Papovavirus warts	Icosahedral Nucleus	45-55	42/72	dsDNA closed circular	–	Polyoma virus of Human wirt virus SV40	Tumours Tumours
2. Adenovirus respiratory	Icosahedral Nucleus	80	252	dsDNA, Linear	–	Human adenovirus	Upper disease
3. Herpesvirus Fever	Icosahedral Nucleus	100	162	dsDNA, Linear	+	Human herpes simplex, types 1 + 2,	Type 1- blisters
Genital herpes							Type 2-
mononucleosis						Epstein-Barr	infectious
chiken-pox						Zoster (varicella)	Singles,
4. PoxVirus	Helical Cytoplasm	160 x 200	–	dsDNA	+ –	Variola (human) Vaccinia (human)	Smallpox Immunity

4. PoxVirus	Helical Cytoplasm	160 x 200	-	dsDNA	+	Variola (human) Vaccinia (human)	Smallpox Immunity to small pox
5. Parvoviruses	Icosahedral Nucleus	18-24		12/32 ssDNA (+/-)	+	AAV Adeno associated virus	
6. Reovirus	Icosahedral	75-80		180/ dsRNA 270 10 or more segments	–	Reovirus of human	Respiratory and GI tract illness
7. Picorna virus	Icosahedral	18-30		32 dsRNA (+)	–	Poliovirus Coxackle Rhinovirus Foot and mouth disease virus	Poliomyelitis Myositis Common caid Cattle disease
8. Togavirus (arbovirus)	Icosahedral Cytoplasm	-50		ssRNA (+)	+	Type A (Alphavirus) Type B (Flavi virus)	Encephalitis encephalitia
9. Rhabdovirus	Helical Cytoplasm	70 x 175	–	ssRNA (-)	+	VSV Rabies virus	Infects cattle Rabies
Paramyxo virus	Helical Cytoplasm	-50	–	ssRNA (-)	+	Mumps virus Mealses virus	Mumps
Orthomyxo virus	Helical Cytoplasm	18-2	–	ssRNA (-)	+	Influenca A, B and C segmented	Influenza
Bunya virus	Helical	90-100	–	RNA (-)	+	Bunyawera viruses segmented	Encephalitis
Coronovirus	Helical Cytoplasm	80-120	–	Segmented RNA	+	Human strains	Common cold, GI disease
Arenavirus	Helical	100-130	–	RNA	+	Lassa virus	Generalized infection rare, serious
Retrovirus	Helical	-100	–	2 dsRNA + 75 & 4S host RNA	+ +	RVS Mouse viruses	Sarcoma Leukemia in mice

5

MYCOPLASMA

The discovery of Mycoplasma goes back to the 19th century (1898), when two French scientists studying the pleural fluids of cattles suffering from a disease called bovine pleuropneumonia, discovered some microorganisms that were totally different from the ones known before. The disease was highly contagious and became widespread in Europe during the 19th century. The disease causing organisms were different from bacteria and were given the name pleuropneumornia like organisms (PPLO). These PPLO were aerobic and could be cultured on a rich organic medium and were given the name, *Mycoplasma mycoides.* The first cultured forms were some times spheroidal in shape, but also produced a wide variety of minute granular structures, thin branching filaments, asteroid structures or many irregular forms. Due to this, the mycoplasma were called pleomorphic.

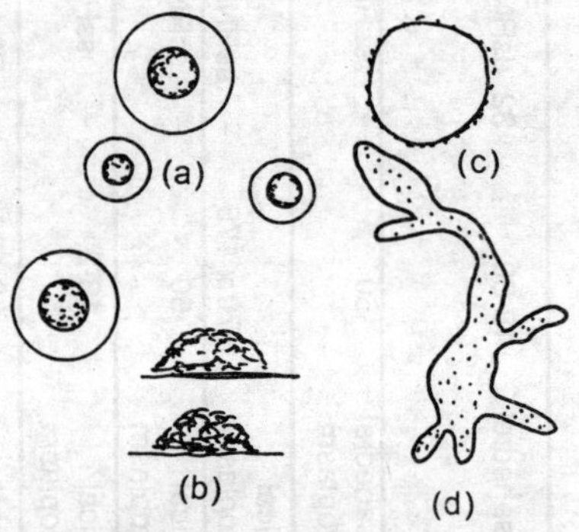

Fig. 5.1 Mycoplasma

Colony morphology and cell shapes of a mycoplasma. (a) entire colonies (b) longitudinal section of two colonies growing on agar surface (c) spherical form (d) irregular filamentous form.

After the initial discovery of PPLO from bovine serum, similar pathogenic forms were reported from a wide variety of veterinary sources like sheep, goats, dogs, rats and mice. In animals, they (PPLO) are associated with a variety of pathological conditions such as, rheumatic or arthritic diseases, infections of mammary glands, respiratory tract and adjacent tissues and inflammation of urinogenitial system.

Some of the PPLO are also known to be human pathogens. The human mycoplasms (PPLO) flora consists of *M.pneumoniae* causing primary atypical (atypical pneumonia is not attributed to any bacteria that generally cause pneumonia) Pneumonia.

Subsequent researches (Doi et. al, 1967) on mycoplasma indicated that they also cause disease to plants. Plant diseases of mycoplasma origin are–aster yelllow, potato witches broom, dwarf disease of mulberry, spike disease of sandal etc.,

Many mycoplasmas are also found in saprophytic conditions in decaying organic matter, sewage etc.

GENERAL PROPERTIES OF MYCOPLASMA (PPLO)

The following are the general properties of mycoplasma.

1. They are the smallest, free living organisms (about 125-250 nm in diameter).
2. They can easily pass thorough bacterial filters.
3. They are highly pleomorphic – branched, beaded, swollen, ringed or even stellate. They may even be spherical, ovoid, or filamentous (branched). Sometimes they resemble the hyphae of fungi in their morphology.
4. They lack a cell wall. The outer most boundary is a three layered plasma membrane.
5. They are either parasitic or saprophytic.
6. They can be cultured in a cell free medium with an abundance of organic matter.
7. They are aerobic as well as anaerobic.
8. Some members (Mycoplasma) require sterol for growth, while others (Acholeplasma) do not require it.
9. Some mycoplasma called 'T' strains mycoplasma require urea and require a lower pH for growth than others.
10. Ribosomes are present in the cell, but different from those of true bacteria.
11. Mycoplasmas are resistant to penicillin but sensitive to tetracyclins.
12. Specific details of reproduction are not known. Reproduction may take place by elementary bodies budding or binary fission.

13. Mycoplasmas are gram negative.
14. Spores, flagella etc. are absent.

Classification :

The original name for mycoplasma was PPLO, as they resembled the pleuropneumonia agent in some respects. Their varied morpholigical nature, spheroidal shape, or occasional branched filamentous structure and mold like appearance indicated that they have affinities with the group Actinomycetales, that includes mold like bacteria. However, their taxonomic position was much under debate. They were also classified under class Schizomycetes (Bacteria), and order Mycoplasmatales, family Mycoplasmataceae and genus Mycoplasma.

Since doubts persisted about the taxomic affinities of mycoplasma with bacteria, in 1966, the International Committee on the bacterial nomenclature constituted a subcommittee to go into all aspects of mycoplasma systematics and offer suitable recommendations. Accordingly, the subcommittee deliberated and gave the following suggestions.

(i) The mycoplasma are not to be regarded as stabilized L–L forms of bacteria (L forms of bacteria do not have cell walls like mycoplasma).

(ii) Considering their unique characters as different from bacteria and viruses, the mycoplasms should be accorded the rank of a class – Mollicutes along with bacteria (Schizomycetes).

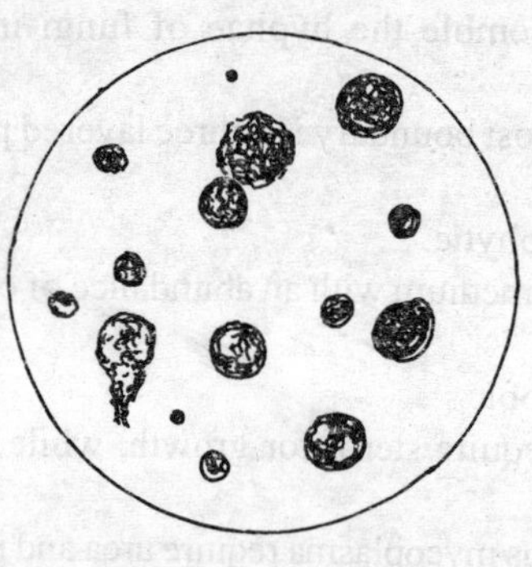

Fig. 5.2 Mycoplasma
Fried egg appearence of Mycoplasma colony under microscope

The class Mollicutes is divided into two orders – Mycoplasmatales and Acholeplasmatales each with a single family – Mycoplasmataceae and Acholeplasmataceae. There are only two genera, one belonging to each family. These are mycoplasma and Acholeplasma (Edward and Freundt, 1970). In addition to these two genera, two more genera of uncertain affinities are also included. These are – *Spiroplasma* and *Thermoplasma.*

The main criterion for the distinction between Mycoplasmataceae and Acholeplasmataceae is the requirement of sterol for growth in the former, and its non requirement in the latter.

The mycoplasma are also classified into three groups based on their physiological properties. These are :

(i) Carbohydrate fermenting – non sterol requiring
(ii) Carbohydrate fermenting – sterol requiring
(iii) Non carbohydrate fermenting – non sterol requiring.

In the groups i and ii above, acid is produced from sugars such as glucose, fructose, mannose etc. The non fermenting species utilize amino acids and fatty acids as carbon and energy sources (eg. *Mycoplasma arthritidis).*

General characters of families

Mycoplasmataceae

1. They can grow only when sterol is added to the medium.
2. They can grow in a temperature range of 22°C to 41°C with the optimum being 36°C to 37°C.
3. A pH of 7.0 is ideal for their growth.
4. The GC content of DNA is 23-40%.
5. They are mostly animal pathogens (including human beings), rarely plant pathogens.

Acholeplasmataceae

1. Members saprpophytic, present in sewage; sterol not required for growth.
2. Require a temperature range between 37°C to 65°C with optimum being 59°C.
3. Acidic pH range is ideal (1-2).
4. They form typical "fried egg" colonies on the medium.
5. The GC content of DNA is 25-26%.

Cell Structure (*Morphology and fine structure*)

Mycoplasma cell cultures show short or long branching filaments with many spherical coccal bodies. The morphology of the cell depends on the nutritive properties of the growth medium particularly the ratio of saturated fatty acids to unsaturated ones. This factor particularly affects the membrane structure of the organisms.

The size of the cell is about 125-250nm. It is from cells of this size which is minimal to the existence of the organism further growth begins. Some mycoplasma may reach a size of 0.5 to 1.0 μm.

The cell has *no cell wall* and there are no locomotor organs. The lack of cell wall is responsible for its fragility and plasticity with which it can assume various shapes. The outermost boundary of the cell is a plasma membrane which is about 8-15 nm in thickness. The membrane is triple layered with two electron dense layers separated by a translucent layer. In some species, like *Mycoplasma pulmonis, M. gallispacum* etc surface spikes are present in the membrane resembling *Myxoviruses.* The electron translucent layer is made up of long chain of fatty acids while the electron dense part is made up of proteins and carbohydrates. Unlike in the bacterial cell, the lipids present in the membrane are either cholesterol or carotenol. Some of the lipids are glycolipids and the sugar moities of these molecules are important serologically as determinants of antigen reaction in fixing and neutralizing the antibody reactions.

The membrane polyopeptide chain (protein), which is either external or interwoven also contributes to the antigen reaction.

The absence of a protective cell wall imposes a special responsibility the cell membrane to maintain the integrity in the face of varying external osmotic pressures. Cholesterol interspersed with phospholipids is known to play a role in the maintenance of membrane integrity. The absence of cell wall offers immunity to mycoplasmas from the action of antibiotics like pencillin and cycloserine which act on the process of cell wall synthesis.

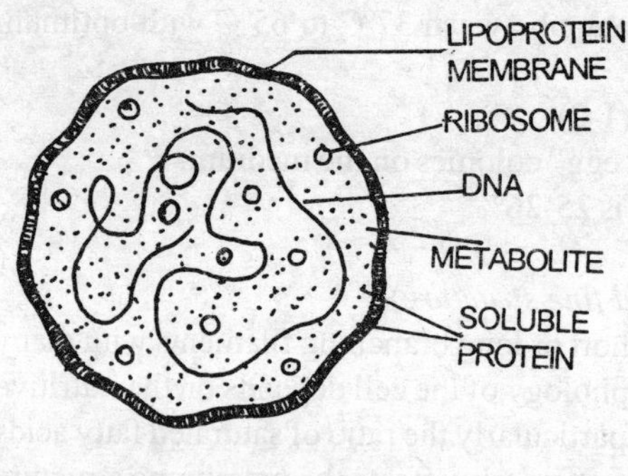

Fig. 5.3 Mycoplasma
A profile of mycoplasma cell

The mycoplasma membrane is also the site of many metabolic reactions involving enzymes (which are membrane bound), and transport mechanisms. In *Mycoplasma pneumoniae,* there is a specialized area on the membrane called the 'foot' which attaches to the animal cell surface. A needle like structure is inserted into the cell through which mycoplasma enzymes enter and digest the host cell metabolites.

The cytoplasm of the mycoplasma does not have an internal membrane system *i.e.,* there is no endoplasmic reticulum. There is no well defined mesosome

unlike in bacteria. In *M. gallisepticum,* membrane bound *vesicles* or *blembs* are found whose function is not yet clear. The cytoplasm has only two organelles–ribosomes and nuclear material.
The ribosomes are of the 70s variety and as in bacteria, protein synthesis in these is affected by antibiotics like puromycin, streptomycin, erythromycin, chloramphenicol etc. This aspect is of a considerable significance in the rapeutics. It is worth noting that cyclohexamide, an inhibitor of protein synthesis in 80s ribosomes, has no effect on mycoplasma.

The nuclear material is fibrillary in form and is either centrally placed or dispersed. The DNA is naked and has no protein associated with it. The total content of DNA in a mycoplasma cell amounts to between 1.5–4% of the dry weight of the cell. Circular double standard DNA has been extracted from six species of mycoplasma. Molecular wt. ranges from 444 x10^6 to 1200 x 10^6 daltons. The GC (Guanine/Cytosine) ratio varies between 24–41% in different species. In some cases G/C ratio is used as a taxonomic criterion.

Reproduction

No sexual reproduction has been reported. *It is doubtful whether mycoplasma can undergo congugation of the bacterial kind as it involves the cell wall.* A sexual reproduction is by means of simple fission or formation of elementary bodies.

Very minute bodies called elementary bodies or minimum reproductive units are commonly formed inside the adult cells of mycoplasma. These elementary bodies range in size from 100 - 300μm. They can pass through minute bacterial filters but are viable on culture media. These minute bodies are often mentioned as examples of minutest independently living (viruses are not independently living) entities. The elementary bodies represent the young stage in the life cycle of mycoplasma. They enlarge to form long filaments, spheroidal cells, chains of bead like cells etc. The elementary bodies are released from the larger cells by the rupture of the outer membrane.

There are also reports, of binary fission and budding in mycoplasma, as in yeats.

Culture of Mycoplasma

Mycoplasma can be cultured in the laboratory using various media. The choice of a medium depends on the species to be cultured. Mycoplasma are cultured best on solid or semi solid agar media with a rich meat base and a high supplement of animal protein. They (mycoplasma) do not seem to prefer liquid media, as the growth of the colonies is not profuse on liquid media. PH-of 7.8 is ideal for the growth of cells. Yeast extract (25%), is another useful ingredient to be added to the medium especially to culture animal pathogens. Serum (25%), is another important ingredient in the medium. Some of the 'T' strain mycoplasmas require urea. Either this can be directly added or the 'T' strains obtain it through the serum. Most of the species are aerobes or faculatative anaerobes, but some of them require microaerophilic conditions and CO_2.

Incubation at 37°C is prefered by most of the mycoplasma. But some Acholeplasma grow best at 22-30°C.

The addition of unsaturated fatty acids to the medium encourages growth and for all parasitic species, cholesterol or a related sterol is essential.

Mycoplasma may also be readily cultivated in living chick embryos. Two types of media which are widely used for mycoplasma culture are given below.

1. Mycoplasma Agar
PPLO Agar or BBL mycoplasma agar is to be prepared as per instruction of the brand. Melt the agar and cool it to 45°C; to 7 parts of this add 2 parts of untreated horse serum and one part of 25% yeast extract. The PH should be between 7.6–78.

2. Mycoplasma broth
PPLO broth or beef heart infusion broth is taken and to this is added 10% yeast extract solution and 20% horse serum. PH is adjusted to 7.6-7.8

Isolation and identification of mycoplasma from cultures.
The sample (source) containing mycoplasmas are to be filtered through a membrane filter (pore size 200-450nm). Most mycoplasma and viruses will pass through the filter while other micro organisms are held back.

The filterate should be taken in a platinum loop (previously sterilized by flame) and streaked on the prepared mycoplasma agar. Only mycoplasma will grow on the agar as it is cell free. Even though viruses may be there, they cannot form plaques as there are no living hosts.

Detection of mycoplasma
Mycoplasmas can be identified in the culture by the following three methods– ***Colony formation, Staining*** and ***antigenic*** analysis.

Colony formation

Colony features of mycoplasma are highly specific, and these are used to identify and characterize them. Minute colonies are formed on the agar medium by mycoplasma. After the initial incubation, colonies start appearing within two days and in about a week's time the colonies attain a size of 0.5–1.0mm. Colonies may be observed visually under a dissection microscope.

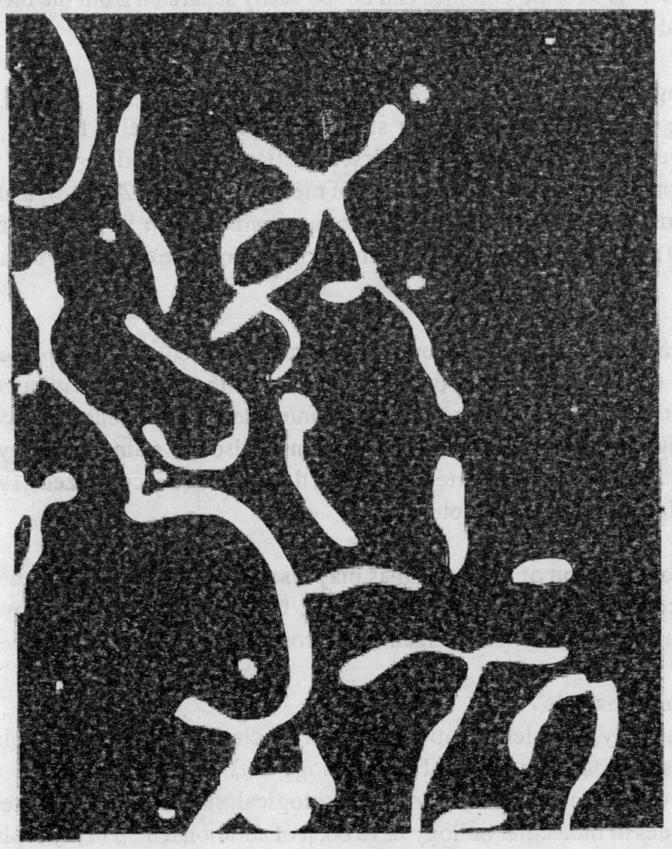

Fig. 5.4 Mycoplasma

Mycoplasma mycoides, gold shadowed electron micrograph, to show branching nature of a filamentous strain

Colonies appear uniformly granular with or without a dense centre. The colonies also have spreading borders showing the typical "fried egg" appearance. The nature colony shows a dark central portion surrounded by pleomorphic or spherical bodies. While the fried egg appearance is most characteristic of mycoplasma colonies, some may not exhibit this character.

Another characteristic feature of mycoplasma colonies is the embedded centre i.e, when the colonies begin to grow, the cell are drawn into the spaces between agar gel, and the colonies form a ball like structure and spread outwards. As a result, colonies can not be easily separated from the medium by usual microbiological practices.

Staining

Aniline dyes are most suited for staining mycoplasma cells. Dienes' stain (discovered by dienes, one of the first investigator of L bodies. Dines' stain can be prepared by dissolving 2.5 gm of methylene blue. 1.25gm. of azure II, 10.0 gms of maltose and 0.25gm of sodium carbonate in 100ml of distilled water. This is the stock solution from which dilutions are to be used for actual stainig.

Cut a small block of the agar containing mycoplasma colonies and place it on a slide. Gently apply a few drops of stain on to the colonies with a cotton applicator. Most of the mycoplasma colonies stain with a dense blue centre and a light blue periphery. Bacterial colonies also are stained initially, but within about 15-30 minutes they get destained (de colorized), while mycoplasmas retain the colour.

Visual observation of mycoplasmas may also be conducted with the help of phase contrast, or dark field microscope. Ultrastructural details, however become obvious only under electron microscope.

Antigenic analysis

Until recently, the identification and species classification of mycoplasma was based mostly on colonial morphology and biochemical reactions. In recent years, however, systematic serological studies supplemented by techniques of molecular biology have been of additional help in mycoplasma identification. In stable suspensions, agglutination has been used for antigenic classification, and for measuring antibodies in naturally and experimentally infected hosts. Mycoplasmas differ from a great majority of bacteria in that their growth on solid media is inhibited by potent antisera.

Immunofluroscence of colonies of mycoplasmas also yield species specific results, and are particularly useful in detecting mixtures of mycoplasma species.

Seriological techniques, however, are complicated because of the tendency of mycoplasma to absorb and hold on to denatured globulin and other constituents from the serum rich medium in which they are grown.

Some of the more recent molecular biological techniques include comparision of G/C ratios of unknown with the established species. Similarly, DNA hybridization between the known and the unknown is another useful method of identification. Another molecular biological technique is to compare the protein banding on polyacrylamide gel between species. Classification and identification of species by this method correlates well with the results from nucleic and hybridization and antigenic analysis.

Mycoplasma and L forms

Under various cultural conditions, such as the presence of antibiotics like pencillin, cycloserine, bacitracin etc. cell wall synthesis in bacteria is affected. Either the bacterial colonies die, or as it happens is some instances (with suitable osmotic composition of the medium) they survive in the form of spheroplasts or protoplasts with modified cell wall or no cell wall. They multiply as variants of parent bacterium and survive. Such wallless forms of bacteria are called L forms (L phase). They may be stable (do no revert to normal from), or unstable (revert to normal wall forms under suitable conditions).

The colonies of L forms of bacteria have a striking similarity with mycoplasma colonies possibly because both of them lack cell wall. Originaly mycoplasmas were thought to be L forms of bacteria. There are, however, a number of differences between the two.

	L. forms	Mycoplasma
1.	Minimal reproductive unit larger	Minimum reproductive unit smaller
2.	Filamentous forms are rare.	Filamentous forms are frequently seen.
3.	Retain in the biochemical machinery required for wall synthesis at least in part.	No wall.
4.	Cholesterol is not a must for growth	Cholesterol required (except Acholeplasma
5.	Normally non pathogenic	Pathogenic (not – all)
6.	Nuclei and base ratio resembles that of bacteria	Does not resemble

Economic importance

Mycoplasma are undoubtedly, of great concern to human beings, as they are known to cause many human, animal and plant diseases.

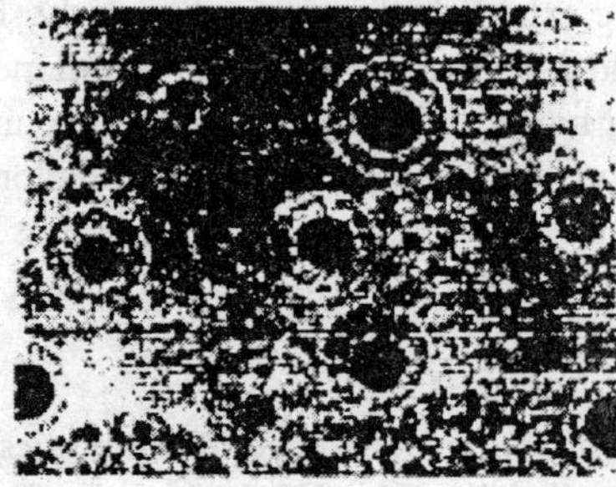

Fig. 5.5 Mycoplasma
Colonies in agar of Mycoplasma homintis. Photographed unstrained, with the low power of the light microscope.

Human mycoplasma

The human mycoplasma consists of about 8 or 9 species. Some of these are – *M. hominis, M. Salivarium, M.Orale, M.Fermantans, M.Pneumoniae, M.Primatum,* T. straits etc. Of all these *M. Pneumoniae* is the most important, while *M. hominis* may function as a low grade pathogen in certain circumstances. *M. Pneumoniae* causes atypical pneumonia. This was first discoveredin 1961, by Maromion and Goodburn. Clinically, *M.Pneumoniae* exhibits symptoms of febrile bronchitis, myalgia, headache, sore throat and cough.

Human infections from *M.Pneumoniae* have been reported from northern temperate zones and Australia. While atypical pneumonia affects algae groups; those between 5-15 years are specially vulnerable.

Clinically, *M.Pneumoniae* can be identified by the culture of throat swabs. Treatment for the disease involves treatment for 2 – 3 weeks wither with tetracyclin or erythromycin.

M.Hominis and 'T' strains of mycoplasma are known to be involved in urinogenital infections, even though reports are not very confirmatory. In some cases atleast, ovarian abscess, salpingitis and haemorrhagic cystitis are traceable to *M.Hominis.*

Animal Mycoplasma

Bovine pleuropneumonia was the first mycoplasma disease discovered. Since then PPLO (mycoplasma), is known to cause a number of diseases in goat, sheep, rat, mice etc. Respiratory tracts, mammary glands and genital tracts are infected.

Treatment involves the use of suitable antibiotics like tetracyclin erythromycin etc.

Plant Mycoplasma

A large number of plants harbour mycoplasmas. They (mycoplasmas) have been detected in electron micrographs of plants suffering from corn stunt, aster yellow and other diseases. Leaf hoppers are known to act as vectors transmitting mycoplasma from the infected to the healthy ones. Many plant disease which were thought to be viral in origin have now been confirmed to be caused by mycoplasma. Some of the important plant diseases caused by mycoplasma are –

1. Aster hellow
2. Potro witches broom
3. Sandal spike
4. Corn stunt
5. Dwarf disease of mulberry
6. Blue berry stunt
7. Little leaf of Brinjal
8. Western x disease

Among the diseases mentioned above, 'Sandal Spike' is the most important, as the sandal tree is one of the most valuable of our forest trees. The heart wood of this semiparasitic tree yeilds the world famous essential oil – sandalwood oil.

The disease symptoms caused by mycoplasma in the sandal tree are of two types–*witches' broom* and *pendulous* spike.

In witches' broom, the leaf area is drastically reduced, branches get stunted and appear like a broom. In pendulous spike, the infected shoot becomes unusually long due to the overactivity of the apical bud and suppression of axillary buds. The affected shoots droop down and hence the name, pendulous spike.

The pathogen (mycoplasma) is transmitted from the infected to healthy plants through insect vectors, such as *Moonia albimalenta, Jassus indicus* etc.

Disease control has not been achieved fully. The Forest Research laboratory at Bangalore has been engaged in the study and control of this disease since a long time. Treatment includes combined therapy of fungicides and antibiotics (Benlate, terramycin or Ledermycin). A wound is made in the affected tree at the base of the stem and a paste of medicine (antibiotic + fungicide) is applied, and the bark replaced over the wound and tied with an adhesive tape. Several treatments of this kind with a gap of not more than 100 days are required, to bring about relief.

Mycoplasma viruses

A number of viruses are known to infect mycoplasma. These are called *mycoplasma phages.* The first mycoplasma phage was discovered in 1970 by Gourlay, and since then more than 75 viruses of mycoplasma have been reported. Three categories of mycoplasma phages have been recognised. These are – Naked bullet shaped particles, spherical, sheathed particles and polyhedral particles with tails. Most of the mycoplasma phages have been obtained from *Acholeplasma laidlawii*. All these are DNA viruses. In bullet shaped viruses, DNA is single stranded. In the other two, DNA is double stranded.

6

BACTERIA

Bacteria are one of the smallest among the microorganisms. Though small in size. They are by no means so, in their significance. With no exception, one can say that human life is being influenced every minute either for good or for bad by these inhabitants of microcosm. Bacteria are highly potent individuals. From the point of view of human welfare, They are of greatest signifiance. In the form of such deadely diseases like cholera, influenza, tetanus, leprosy, tuberculosis, typhoid etc., to name only a few, they are man's worst enemies. They are man's best friends also, in various industries (ex. Preparation ofvarous types of alcohols and acids, curing of coffee, tea and tobacco) and in various other aspects. When such interests are involved, no doubt a separate branch of science 'Bacteriology' is set apart for the study of these.

Historical account:

Credit of discovering Bacteria goes to Antony Von Leeuwenhoek (1676), a Dutch scientist. He was interested in grinding lenses and constructing crude microscopes, with which he used to observe drops of pond water, cells of retina, muscle fibres, nerve cells etc. One day, while he was observing the outer coat of pepper grain with the idea of finding out the reason for the biting taste, he found certain tiny organisms moving to and fro in the fiedl of the microscope. As these were very different from the ones observed before, he made a study of them and called them animalcules, which were later named 'Bacteria'. The next name to be remembered in the history of Bacteria is of Louis Pasteur. Very often and rightly so he is referred as the "father of Bacteriology". He was the one, who found out that Bacteria bring about fermentation. He conducted extensive studies on Bacteria.

Robert Koch, a German Bacteriologist discovered that tuberculosis is caused by Bacteria.

T.J. Burill (1879), an American scientist discovered that Bacteri are responsible for many plant diseases also.

Distribution

With one word perhaps one can explain the distribution of Bacteria ie., 'omnipresent'. There is virtually no place on earth which is free of bacteria. They are present in air, soil, milk, water, in temperate regions, tropical regions. Arctic regions (in snow), in hot springs, in plants, animals not excluding human beings. Bacteria can withstand extreme temperature ranges. For ex. Some can withstand 190°C cooling and heathing upto 78°C.

Morphology

Bacteria are unicellular organisms. On an average, a cell ranges from 0.1 micron to 1.0 micron in size. Exceptionally some forms may be as large as 15 microns (lmicron=l/10,000ofmm).

Some interesting calculations have been made to impress as to how small the bacteria are. Just a drop of water may contain several millions of bacteria. If a Coccus measuring I micron across is enlarged 500 times it would be 0.2" and an average human being enlarged on the same scale would be 2750 feet tall and more than a thousand feet broad at his shoulders. Nearly 500,000,000,000 bacteria would weigh just one gram.

Based on the cell shapes, four morphological forms have been identified in bacteria viz.

1. Bacillus- the cells are rod shaped and elongated. These are the second largest among bacteria eg. Mycobacterium.

2. Coccus- the cells are spherical. These are the smallest among bacteria. Eg. Streptococcus, staphylococcus etc.

3. Spirillum- the cell will be in the form of a loose spiral. These forms are the largest. Eg. Rhodospirillum.

4. *Comma*- the cell will be in the form of punctuation mark, comma eg. Vibrio. Coccus form exists in various ways. When existing as single isolated cell, it is called Micrococcus, in pairs- *Diplococcus*, in chains- *Streptococcus*, in sheets Staphyloccus etc.

Many bacteria exhibit several shapes or forms. This is known as "pleomorphism" (Eg. *Azatobacter; hizobium* etc).

Locomotion

Except the coccus, all the others forms of bacteria may posses organs of locomotion ie. Flagella. The flagellum is a long thread like structure whose length usually exceeds that of the cell. It arises from a basal granule.

On the basis of the number and position of flagella, bacterial forms are classified into-

1. *Atrichous*- no flagella eg. *Cocci*
2. *Monotrichous*- single flagellum present at one end eg. *Vibrio*

3. *Lophotrichous*- many flagella present at one pole eg. *Spirillum undula*

4. *Amphitrichous*- two tufts of flagella attached at either poles of the cell eg. *Spirilla*.

5. Peritrichous- flagella are present all round eg. *Salmonella typhosa*.

Bacteria move with varied speeds. Some can move about 2,000 times their size in an hour. Hay bacillus can travel at a speed of 200 microns per second (Knayasi, 1938).

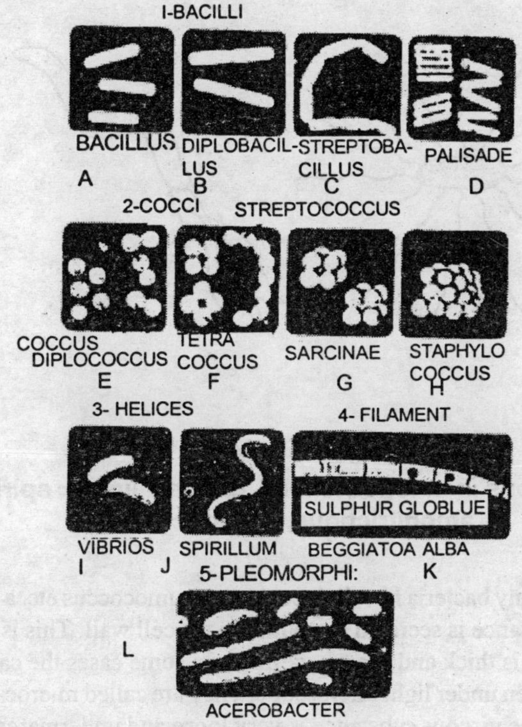

Fig 6.1 Bacteria
Types of morphological forms

Ultra structure ofa bacterial cell:

Electron microscopic studies have revealed the ultra structure of the bacterial cell. The outer envelop is the cell wall internal to which is the cell membrane. Sometimes external to the cell wall there will be aloose slimy layer or capsule.

mesosomes,fat globules, vacuoles, inclusion bodies and nuclear material. We shall study each one of them in detail.

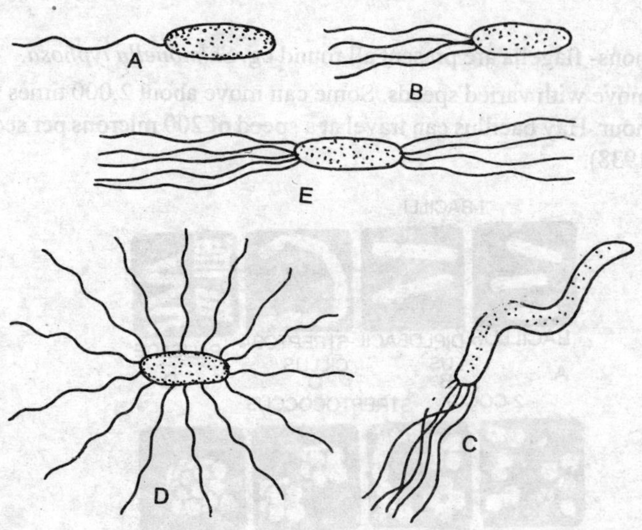

Fig 6.2 Bacteria

a, Monotrichous ,b-c. lohotrichous (b= basillus, c= spirillum), d. amphitrichous, e peritrichous

Capsule : In many bacteria like diplococcus, pneumococcus etc. a viscous or gelatinous substance is secreted surrounding the cell wall. This is called the capsule when it is thick and sharply defined. In some cases the capsules are too thin to be seen under light microscope. They are called microcapsules. In Leuconostoc, the viscous substance is very loose and undermarcated and is known as slime layer. Chemically the capsules are made up of proteins, polysaccharides and lipids. Some slime layer polypeptidesare built with a single amino acid. The capsules may not be present in all the bacteria. In one and the same species (*Diplococcus pnuemoniae*) capsule may be present in one strain ai 1 absent in the other. Usually virulence (Pathogenicity) is associated with the capsule.

Cell wall: the cell wall is thin about 10-25 mm thick and provides rigidity to the cell. Cell wall is visible only under the electron microscope. The cell wall accounts for 20-30% of the dry weight of the cell.

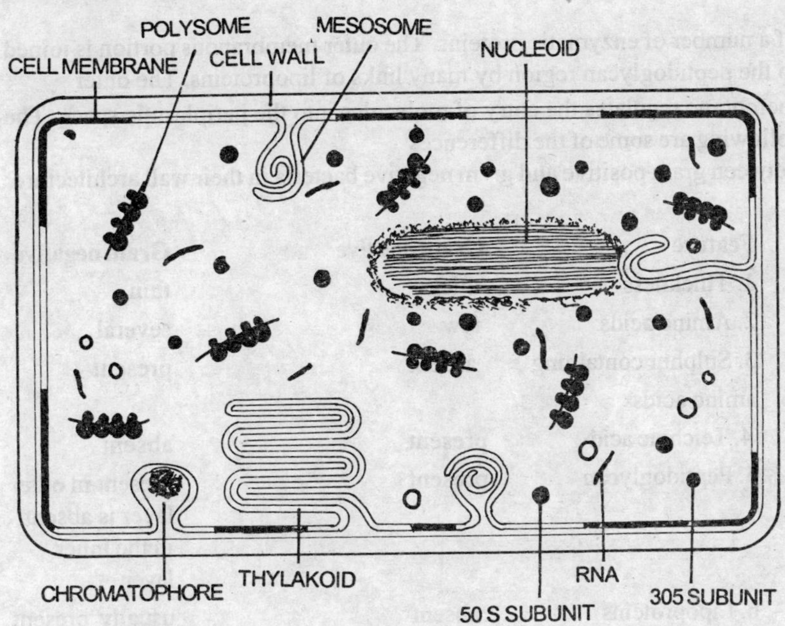

Fig 6. 3 Baacteria

Detailed cell structure (Electron micrographic view)

Chemically, the cell wall consists ofmucopeptides made up of alternating chains of N-acetyiglucosamine and N-acetyl muramic acid molecules. These chains are cross linked by peptide chains. The narrow spaces left by these chains are filled by various other chemical components which vary with the species.

The cell wall composition differs between gram positive and gram negative bacteria. The cell walls of gram positive bacteria contain up to 95% peptidoglycan and about 10% teichoic acids. In gram negative bacteria, the cell wall composition is more complex. It is made up of several layers. Next to the cell membrane is The peptidoglycan layer. Next to this extreroallyisthe periplasmic region consisting

PENTAPEPTIDE BRIDGE
N-ACETYL GLUCOSSAMINE
N- ACETYL NUMERIC ACID
PEPTIDE CHAIN

Fig 6. 4 Bacteria

Chemical structure of bacterial cell wall

of a number of enzymatic proteins. The outer membranous portion is joined to the peptidoglycan region by many links of lipoproteins. The outer membrane regulates the entry of molecules into the periplasmic space. The following are some of the differences
between gram positive and gr-m negative bacteria in their wall architecture.

Feature	Gram positive	Gram negative
1. Thickness	thick	thin
2. Amino acids	few	several
3. Sulphur containing amino acids	absent	present
4. Teichoic acid	present	absent
5. Peptidoglycan	present	present in outer layer is absent in the inner layer
6. Lipoproteins	absent	usually present
7. Lipopolysaccharide	absent	absent
8. Polysaccharide	present	absent

Cell membrane: Internal to the cll wall is the cell membrane or the plasma membrane. It forms the outer boundary of the cytoplasm and is selectively permeable and it thus regulates the entry and exit of molecules into cytoplasm. The membrane is chemically made up oflipoproteins (70:30 protein and lipid) and practically no carbohydrates. Electron microscopic studies (Costerton *et al* 1974) have shown that the membrane is a three layered one with a unit membrane

Structure. There are two electron dense layers encompassing a central electron transparent layer. The outer layers are about 3.5nm thick and the middle electron transparent layer is about 5nm thick. Lipids found in the membrane are phospholipids such as phosphatidyl ethanol amine. The bacterial cell membrane performs the following functions.

1 .Functions as an osmotic barrier

2. Contains enzyme systems involved in biosynthesis of membrane and cell wall components.

3. Contains components of energy generation systems.

Mesosomes:

In some bacteria, particularly in gram positive bacteria, the cell membrane forms vesicles or packet like infoldings into cytoplasm at several regions.

These are called mesomsomes. Mesosomes are the reservoirs of respiratory enzymes and are supposed to be analogous to mitochondria ofeukaryotic cells. The presence of more number ofmesosomes particularly during the log phase of the growth when respiratory activity is high supports the involvement ofmesosomes in respiration. In photosynthetic bacteria, mesosome number is related to pigment concentration and photosynthetic activity. In sporulating bacteria, mesosome formation is a prequisite for spore development. According to peberdy (1980), mesosomes are also invovied in cell wall synthesis and are necessary for cross wall formation during cell division.

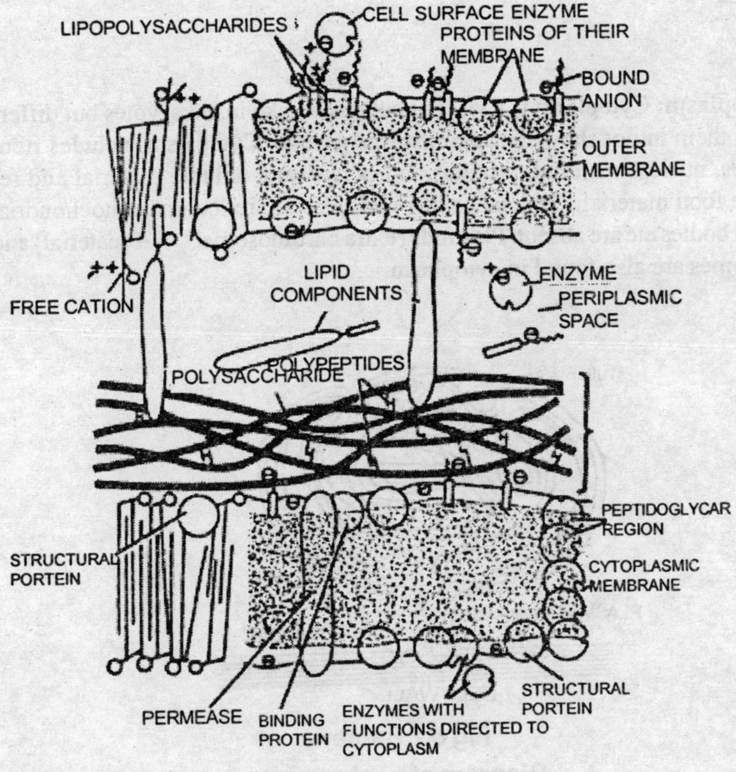

Fig 6.5 Bacteria

Diagramatic represents of wall of Gram -negative bacteria (adopted from Costerton et al 1974)

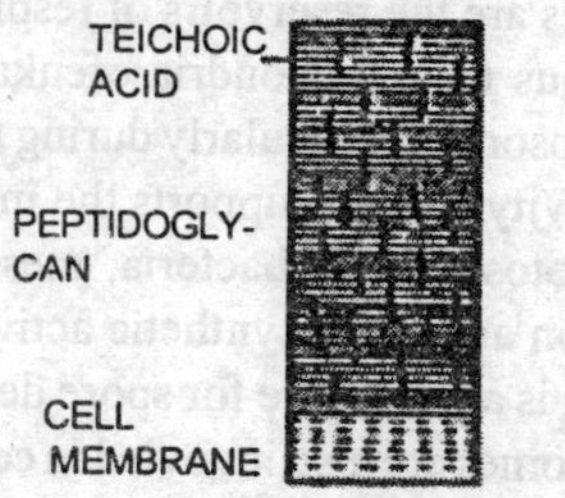

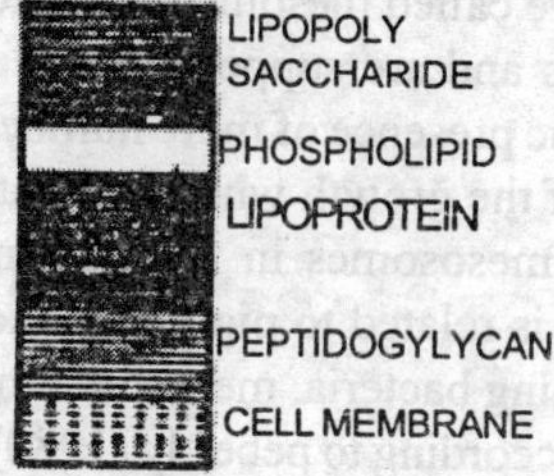

Fig 6.6 Bacteria

A comparison of cell walls of Gram -positive (on left) and Gram -negative (on right) bacteria.

Cytoplasm: Cytoplasm is a viscous substance like in eukaryotes but differs from them in not showing streaming movments. Cytoplasm includes ribosomes, nuclear material, proteins and other water soluble material and reserve food material. Organelles like endoplasmic reticulum, mitochondria, golgi bodies etc are absent. Plasmids (extra chromosomal DNA material) and episomes are also found in cytoplasm.

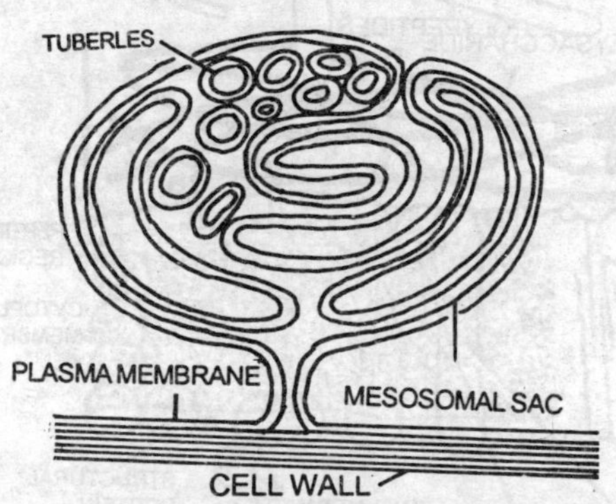

Fig 6.7 Bacteria

Diagram of a mesosome

Ribosomes

They are dispersed throughout the cytoplasm and as in eukaryotes are invoviedin protein sythesis. The number ofribosomes per cell varies and may reach upto 15,000 per cell. Higher number ofribosomes are found during increased activity of protein synthesis.

Each ribosome is a ribonucleo protein particle (60-40 RNA: protein) of about 100-200 A° in diameter. Bacterial ribosomes are of the 70s type as noticed by their sedimentation properties. Each ribosomes is made up of two subunits 50s and 30s. Each 50s sub unit consists of one molecule of 23s r-RNA, one molecule of 5s r-RNA and 32 different kinds of proteins while the 30s sub unit has one molecule of 16s r-RNA and 21 different kinds of proteins. During protein synthesis many bacterial ribosomes are held together by a mRNA chain and are called Polyribosomes or Polysomes.

Nuclear material: Nuclear material is dispersed in the centre. It is called a nucleoid as it has no membrane or nucleolus. Electron micrographs suggest that the nucleoid consists of closely packed fibrillar DNA(Kavenoffand Bowen 1976). Studies indicate that it is a single molecule of double stranded DNA. Some recent works indicate RNA may also be associated with DNA. According to Kieppe, et al (1979), the DNA in *Escherichia* coli is in a supercoiled state in the centre while at the periphery it is loosely spiralled. In E.*coli* the DNA has anaverage length of l000*u*m and contains 5x103 kilobase pairs with amolecular weight of 25x109 daltons. The bacterial DNA molecule is often refered to as the bacterial chromosome.

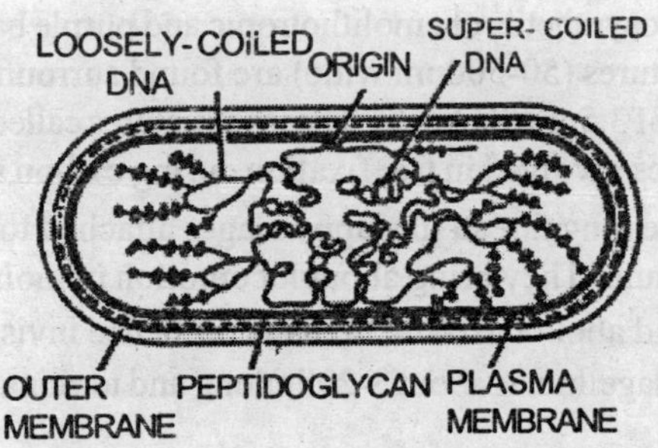

Fig 6.8 Bacteria

A diagramatic model of the organization of the Escherichia coli

Plasmids and episomes: In many of the bacteria, DNA is present outside the chromosome (nucleoid). Such extra chromosomal DNA is often referedto as plasmids or episomes. While it is difficult to dintinguish between the two and they are often regarded as synonymous, plasmids are supposed to be free of the chromosome and episomes may be free or integrated to the chromosome.

Plasmids confer on the bacterial cell, drug resistance, toxin producing ability etc. The size ofplasmid DNA is aoubt one tenth of nucleoidDNA but a bacterial cell may have one to many plasmids. Plasmid DNA is also double stranded and circular and it is also not having any membranous envelop.(see later for details)

Cellular reserve materials : Avariety of reserve materials are found in the bacterial cells. Of these starch, glycogen and poly and hydroxy butyric acids are important. In some cases volutin granules are also found. Sulphur bacteria accumulate sulphur transiently during the process of H_2S oxidation. While bacteria do not have any membrane bound organelles, gas vesicles, chlorobium vesicles and carboxysomes bound by non unit membrane have been described in some photosynthetic bacteria.

In the photosynthetic green bacterium *Chlorobium*, the photosynthetic apparatus has a distinct intracellular location and is bound by a series of cigar shaped vesicles arranged in a cortical layer that immediately underline the plasma membrane. These structures are about 50nm wide and 100-150nm long and are enclosed in a single layered 3-5nm thick membrane and contain photosynthetic pigments.

In a number of photosynthetic, chemolithotropic and purple bacteria anumber of polyhderal structures (50-500nm wide) are found surrounded by a single layered membrane of 3.5nm thickness. Theses structures called carboxysomes contain key enzymes involved in Co, fixation during carbon assimilation.

Flagella: These are long thread like appendages attached to the cells found in bacilli and spirillum. They bring about locomotion in moist medium. The flagellum is thin and about 0.02*u*m in diameter hence invisible under light microscope. Each flagellum is about 3-20llmlong and terminates in a squarish tip.

Electron microscopic and X-ray diffraction studies have shown that the flagellum is made up of three parts viz., filament, hook and basal body. Filament is made up ofprotein fibres and about 13-14nm in diameter. It is outside the cell and connects to the hook at the cell surface* Hook is bent and is thicker than the filament (17nm) and it penetrates the cell wall. The basal body is a spindle shaped structure that joins the hook to the cell membrane. A short collar like structure attaches the basal body to the hook. In gram negative bacteria, the basal body has two pairs of rings. The outer pair (L and P rings) is fixed to the cell wall while the inner pair (S and M rings) is attached to the cell membrane. In gram positive bacteria the basal body is much simpler. It has only one ring.

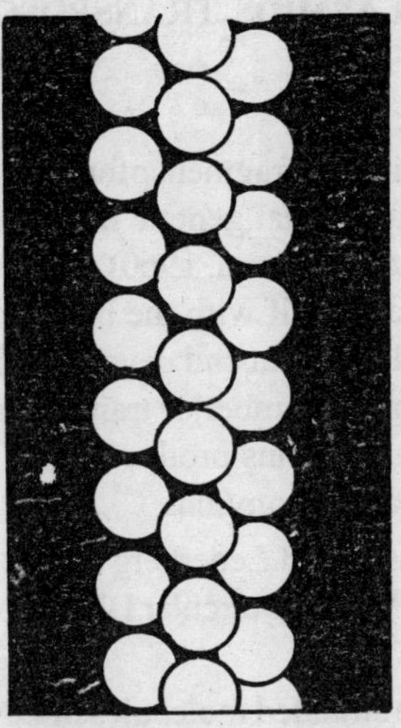

Fig 6.9 Bacteria

A model showing three helically wound strands in a bacterial flagellum

Funbriae(pili):

These are short, very fine, hair like processes found in some gram negative bacilli. Also called pili they are about 0.5lLlm in length and less than IOnm thick. They originate from the plasma membrane and their fucntion is to adhere to the cells particularly during conjugation. Some pili are longer than the rest in some cells. fSuch cells are 'male' and the pili are called sex pili.

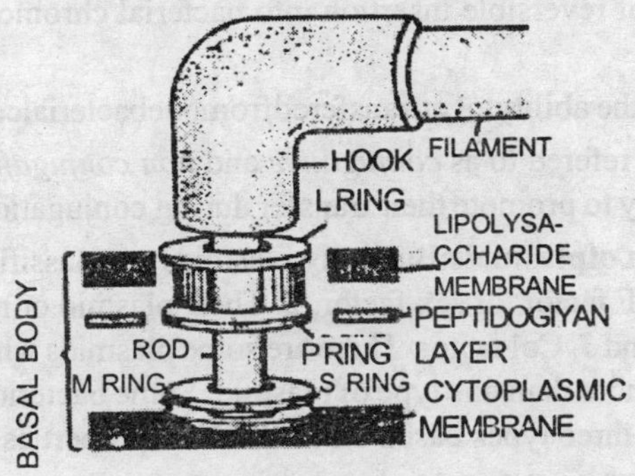

Fig 6.10 Bacteria

Interprective sketch to show the attachement of rings tocell wall and cell membrane of a Gram negative bacterial cell

A BRIEF ACCOUNT OF PLASMIDS, TRANSPOSONS AND DRUG RESISTANCE

Plasmids

The bacterial cell has an additional fragment ofnu^cleci acids in addition to its normal chromosome. This additional genetic material was given the name episome (Jacob, ShcaefferandWollmann, 1960). Episome can exist either independently or it can integrate itself with the bacterial chromosome. Plasmids however are always independent and never integrate with the bacterial chromosome. Plasmids control a number of traits in bacteria such as - toxin production, antibiotic resistance, pilus production etc. Some people regard that episomes and plasmids are synonymus.

The word plasmid was first used by J.Lederberg (1952). However, the term is restricted only to bacterial cells where circular DNA molecules exist independent of the main genome

Plasmids are circular double stranded molecules of DNA and exist independently of chromosomal DNA in bacterial cells. They undergo replication during cell division and are carried to both the daughter cells. The plasmid has a right handed superhelical coil during the resting stage. There, is one super helical turn for every 400-600 base pairs. The twisted conformation is known as **covalently closed circular DNA (CCD).** Cleavage results in the conversion from CCCD to open circular form.

The following are some of the properties of the plasmids.

1. They are made up of DNA (double stranded).
2. They are capable of independent replication
3. Some plasmids control bacterial conjugation
4. Capable of reversible insertion into bacterial chromosome (only some plasmids)
5. They have the abilitytogettransferedfromonebacterialcell to the other.

Plasmids are refered to as *conjugative* and *non conjugafive* based on their ability/inability to promote their transfer during conjugation.

Classification ofplasmids:Basically plasmids are classified into three Categories1. The F factor or sex factor, 2. TheR plasmid or resistance transfer factor (RTF) and 3. Col factor. There are some plasmids which may share the properties of more than one type of plasmid. Some bacteriologists recognise more than the three types based on some other properties.

F factor. The F factor plasmid confers the ability to the bacterial chromosome to get transfered to another cell (recpient). Hence the donor is said to be F+. F factor integrates with the bacterial genome facilitating its transfer. Because of this integration, it is often refered to as an episome.

R. Plasmid. These plasmids, when present confer on the bacterial cell resistance to several antibiotics such as streptomycin, tetracycline, chloramphenicol, sulphonamid etc. R plasmid mediated antibiotic resistance was first discovered in Japan in 1956 when *Shigella* became resistant to several antibiotics at one step. This was due to the transfer of R plasmid from one strain to the other resulting in not only antibiotic resistance but also the ability of a donor (it can donate R factor to other sensitive strains making them resistant).

R plasmid is a small extrachromosomal DNA ring. Sometimes the Rplasmid has two parts - a. Part having drug resistant genes and b. Part having plasmid transfer genes.

The col factor (plasmid). Certain bacteria like *Escherichia coli, Shigella, Salmonella* etc., produce toxins called *colicins*. These are toxic proteins and can kill other bacteria. The ability to produce *colicin* rests in the plasmid called col plasmid or *colicinogenic factor*.Col factors resemble phages since they cause death of bacteria, but they differ from phages as they are not released out during the lysis of the bacterial cell. Like the F factors colplasmids also can be transfered from one bacterial cell to the other.

The number of plasmids per bacterial cell varies. It may be *single copy* or *multiple copies*. Multiple copy plasmids which replicate indepdnely have been used in gene cloning experiments. A plasmid is cleaved and a desired gene is inserted into the plasmid DNA. The plasmid DNA during its replication produces multiple copies of foreign DNA. Such plasmids are called vectors.

Transposable elements

In prokaryotic as well as eukaryotic geriomes certain sequences ofnucleotides are capable of moving from one site to the other and are called Transposable elements. There are three kinds of *Transposable elements*. These are- Insertion sequences (IS elements), transposons (TN elements) and retroelements. Only transposons are described here.

Transposons: Many plasmid genes confering drug resistance are capable of dissociating themselves from the plasmid and transfered to another location ie,, on to a different DNA molecule. From there it can shift to another location. In otherwords a gene sequence can hop from a plasmid to a chromosome to another plasroid and also to a phage genome. Such movable particles of DNA are called transposons. Transposons are designated as Tnl, Tn2, Tn3 etc.

Transposons in prokaryotes. These were first discovered by Hedges and Jacob (1974) for the sequence ofampicillin resistant nucleotides which get transfered from one plasmid to the other thus transfering drug resistance- It was demonstrated that the two ends of the two strands of transposons had complementary nucleotides, but in reverse order. For example if one end had GTCTGGG, the other end had CCCAGAC.

Transposons in eukaryotes. Transposons are also found in eurkaryotes such as *Drosophila*, maize etc, Barbara Mcclintock hasrecevied Nobel prize for her work on transposons which she called *Jumpinggenes*.

Mechanism of transposition. Berg (1977) has proposed a mechanism for the transposition of one of the transposons (Tn5). The Tn5 element determines resistance to the antibiotic Kanamycin and can betransposed from the lambda phage to the Esherichia coli chromosome and from one chromosomalsite to the other.

Transposition consists of three stages - cleavage, realignment of ends andligation. An enzyme complex transposase recognises and binds to speci sequences at the ends ofTn5 element and to another DNA molecule (recipici The enzyme cuts the transposon at either ends and also nicks the recipient DIS After insertion the cut ends are ligated in both the termini.

Uses of Transposons. They can beused as genetic markers, because they change the pattern of cuts by restriction endosnucleases. Strains with different types ol transposons can be easily identified because of the specific drug resistance they carry. Insertion of transposons may be used to induce mutations as has beer shown inmaize, where the transposon acts as a mutator gene inducing the mutable genes. Some recent experiments have shown that transposons can also be used as transformation vectors

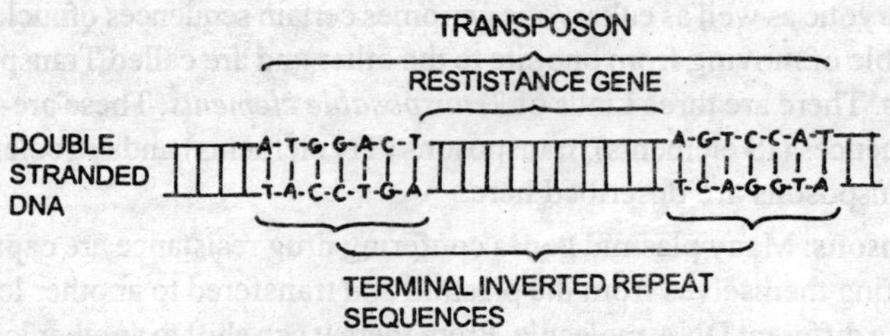

Fig 6.11 Bacteria
Diagramatic representation of a transposon

Durg resistance is a major problem in thechemotherapeutic control ol pathogenic bacteria.

It has already been pointed out that drug resistance is confered on those bacteria which posses **R** plasmids. These are found in both gram positive and gram negative bacteria **R** plasmids consist of two parts-transfer factor (T) and durg resistance factor. T factor is responsible for the independent replication and transfer of the **R** factor. Plasmids carrying **T** factor are called *conjugative R plasmids* while those R plasmids without a T factor are called *non conjugative R plasmids.*

Drug resistance gene sequence can be transfered from plasmids to the bacterial genome. Specific transposons carrying the drug resistant gene transfer them tc differentDNAmolecules.

Mechanism of drug resistance

Davies and Smith (1978) have classified, the various modes of drug resistance mechanism as follows

1. Alteration of target site. In this, the binding of the drug to the target site (in the bacteria) is reduced or removed by the alteration of site, so that the drug can no loger be effective. A specific example is that of erythromycing resistance in *Streptococci* and *Staphylococci.* A plasmid regulated enzyme, methylase causes methylation of 23s r RNA (in the bacteria). This prevents the binding of drug and confers resistance.

Drug	Target	Plasmid enzyme
Erythromycin	50's Ribosomal	
Linomycin	sub unit (235rRNA) methylated; proteins synthesis inhibited	Methylase

2. Blocking of antibiotic transport: Plasmids prevent the entry of antibiotics which are structural analogues of bacterial substrates. Tetracycline resistance is partly due to bloackade of drug transportation.

3 .Enzymatic in activation. Resistance to penicillins and cephalosporms is due to plasmid mediated inactivation chlormaphenicol resistance also is due to acetycation induced by acetyl transferases which again are plasmid regulated.

4. Metabolic change. Change of metabolic pathway that is inhibited by the antibiotic, but producing the same metablite is another method of drug resistance. Resistance to sulphonamides and trimethoprim belongs to this category. Other methods of drug resistance include - a. Saturating the drug by increasing the concentration ofinhitibed enzyme.

b. Production of inhibitor antagonistic metabolite and c. Decreasing the dependence of bacterial cell on the metabolic pathway inhibited by the drug.

PHYSIOLOGY AND NUTRITION: Nutritionally bacteria may be classified into two types. 1. Autotrophic bacteria 2. Heterotrophic bacteria.

Autotrophic bacteria: Autotrophic bacteria prepare their own food. They are of two types I. Photosynthetic bacteria 2. Chemosynthetic bacteria.

Photosynthetic bacteria: On the basis of colour, photosynthetic bacteria fall into two principal groups. The green and the purple bacteria. All the three forms (Rods, Spheres and Spirilla) are found among the photosynthetic group.

The photosynthetic bacteria (like green plants) contain chlorophyll which is chemically very much similar to the chlorophyll of higher plants.

The chlorophyll found in the purple bacteria (Bacterial chlorophyll) differs chemically from green plant chlorophyll in several ways. As a result of this, the pigment will be pale blue or gray. The colour is caused by the presence of various carotenoid pigments.

Chemosynthetic bacteria: These are pigmentless bacteria (but autotrophic) that obtain energy by the oxidation of inorganic compounds. After oxidation, the inoiganic element will be left as residue in the cytoplasm. Depending on the inorganic compound used in oxidation, bacteria may be classified into sulphur bacteria, iron bacteria, hydrogen bacteria, carbon bacteria etc.

One of the sulphur bacteria (Thiobacillus thiooxidans) oxidies elemental sulphur tosulphuricacid.

$S+H_2O+1/2O_2 - H_2SO_4$energy

Another group of sulphur bacteria (Beggiatoa) oxidises hydrogen sulphide to elemental sulphur

$H_2S+1/2O_2-S+H_20$

The oxidation of Iron is carried by Ferrobacilus

Fe^2 $F^4+ +e^-$

the ferric iron is deposited as insoluble Ferric hydroxide.

Heterotrophic bacteria. These are pigmentless bacteria that obtain food either by parasitic mechanism or by saprophytic mechanism.

Parasitic bacteria. They obtain their nutritional requirements by parasitising other organisms both plants and animals. Parasitic bacteria may be either pathogenic or non pathogenic. The pathogenic bacteria secrete toxins which bring about the disease. Not pathogenic bacteria which live in human beings are in fact helpful. For ex. The intestinal bacteria synthesize B vitamin which could be used by human beings.

Saprophytic bacteria: Saprophytic bacteria obtain food anaerobically oxidising organic matter. Depending on the nutritional source they may be classified into I. Fermenting bacteria 2. Putrefying bacteria, Fermenting bacteria. These bacteria use carbohydrates as the respiratory substrate. They anaerobically respire and incompletely oxidise the carbohydrates resulting in the formation,of alcohols. The following is the reaction.

$C_6H_{12}O_6 - 2C_2H_5OH + 2CO_2 + energy$

ethyl alcohol

Putrefying bacteria. These make use of protein or nitrogenous matter as the respiratory substrate resulting in the liberation of ammonia, marsh gas (methane) etc. The evolution of these gases could be commercially exploited.

REPRODUCTION:

Bacteria reproduce asexually as well as sexually. Asexual reproduction takes place by the following methods 1. Fission (cell division) 2. Budding 3. Endospore formation.

Cell division. generally bacterial cells do not divide mitotically as is true for all prokaryotic cells. There are tliree aspects of cell division called binary fission. These are. 1. DNA duplication 2. DNA partitioning 3. Cross wall formation. The cell division is completed by the doulbing of all the cell components and their precise distribution and partitioning between the two daughter cells. Unlike in eukaryotic cells there is no spindle formation and no resolution ofchromatin into chromosomes.

Electron microscopic studies have revealed the following details in bacterial cell division.

DNA duplication. The two strands separate and each strand then replicates a new strand.

DNA partitioning. This involves the distribution of DNA between the twodaughter cells. As the DNA is free and not condensed into chromosomes it is a complicated task to assure equal distribution of genetic material between the two daughter cells. It is believed that (Kleppe et al, 1979) the DNA molecule attaches itself at some point to the plasma membrane and after duplication, one strand swings towards one half of the cell keeping itself in the same attached position. Both the havies of cell receive one double stranded DNA each. The attachment of DNA to the plasma membrane is to assure proper distribution without entanglement.

Cross wall formation. After the DNA has been properly distributed between two halves of the cell, cross wall formation begins. The cell wall projects inwardly mid way between the two nuclear materials that have now become two. Cell wall materials are deposited between the membranes and the cell gets divided into two. It is beleived that mesosome plays an important role in the synthesis of the wall material during cross wall formation. Hydrolase enzyme is known to bring about the separation of two daugther.cells in *Bacillus subtilis and Staphylococcus facalis* (Peberdy , 1980).

Budding. In this process, a protrusion first develops, into which cytoplasm migrates. The nucleus splits into two and one bitenters the protrusion. This is later cut off from the cell by a cross wall. This is the bud. At maturity, this separates from the parental cell and develops into a new individual.

Endospore. Many of the bacteria like *Bacillus, Clostridium* etc., form endospores at certain stages in their life ccle. Strictly endospore formation can not be called reproduction because only one spore is formed per cell. Endospores are highly resistant to environemntal conditions and may be regarded as an attempt by bacteria to withstand the adverse environmental conditions. Endospores ar so highly resistant to desiccation, chemicals, temperature and radiation that they are vialable even for centuries. Some endospores even withstand boiling temperature also for a long time.

As the spores are formed inside the parental bacterial cell, they (spores) are called endospores. On germination each endospore develops into a single bacterial cell. The spore wall assures the endospore high resistance due to the presence of compounds like diplolinic acid, calcium and a very low water content.

Shape aod position of endospore. The spores areof various shapes and occupy various positions in the cell. They may be spherical or oval in shape and in position may be terminal, sub terminal or central. They may be bulging giving a swollen appearance to the cell or non bulging conforming to the outer parameters of the cell.

Ultra structure of endospore. The outermost layer of the endospore is the exosporium which is thick and wrinkled. Chemically it is made up oflipids and proteins with a low content ofmethionine and cysteine. Internal to the exosporium are two spore coats- the outer and the inner. The spore coats may be smooth or show various types of ornamentations like ridges, grooves etc. Chemically the spore coats are made up of lipids, proteins and glycopeptides

Гhe protein of inner spore coat has a high quantity of sulphur containing amir acids. The spore coats surround a space called cortex which can be divided inl outer cortex and inner cortex. Chemically cortex consists of dipicolinic ach peptidoglycan and calcium ions. The cortex is internally lined by the spore wa internal to which lies the cell membrane. Chemically the membrane consists (peptidoglycans but no teichoic acid. The membrane surrounds the central coi of the endospore. The core consists of nuclear material and cytoplasm. Th region has t)NA, RNA and proteins.

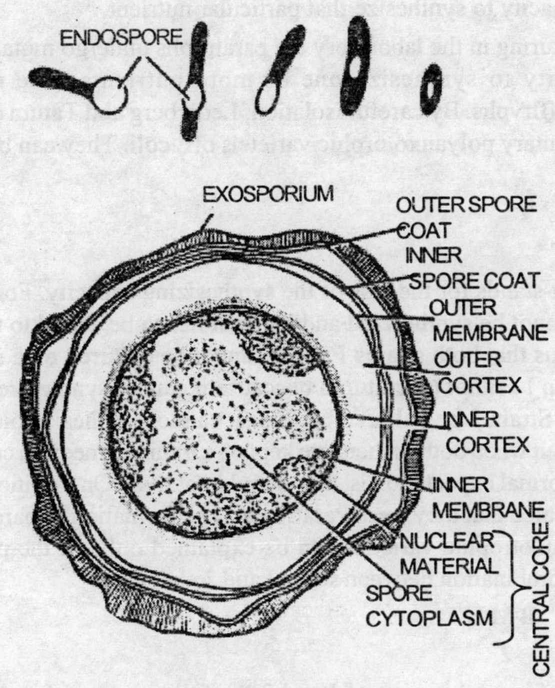

Fig 6.12 Bacterria

A -E shape and position of endospores in bacteria

F. electron micrograph of T.S of an endospore Bacillus cerus

Sexual reproduction

Sexual reproduction as shown by genetic recombination has been demonstrated in bacteria. These are three types. Genetic Recombination,Genetic tration and Transduction.

Genetic recombination- Lederberg and Tatum (1944) demonstrated this in experimenting with the biochemical mutants of *Escherichia coli.* These bacteria require the following four nutrients 1. Biotin 2. Methionine 3. Leucine 4.

Threonine. But they can be grown on a minimal medium (a medium containing only agarand carbohydrates) because, by using carbohydrates they can synthesize all the above four nutrients. They are called Taratrophs' and can begenotypically designated as follows:

$B^+M^+ L^+T^+$

In this, the letter stands for the nutrient (B=biotin etc) and the sign + stands for the capacity to synthesize that particular nutrient.

By subculturing in the laboratory the paratrophs undergo mutation and lose the capacity to synthesize one or more nutrients. The mutants are cal\edAuxfftrvphs. By careful isolation, Lederberg and Tatum obtained two complementary polyauxotorphic varieteis ofE.coli. They can be designated as follows:

1. $B^+ M^+ L^+ T^+$

2.$B^-M^-L^+T^+$

(The sign-stands for the loss of the synthesizing capacity. For ex:L'means leucine cannot be synthesized and hence it has to be added to the culture.). It is obvious that both strains I aod 2 cannot be cultured on a minimal medium. Stran I cannot be cultured beccause it cannot synthesize leucine and theronine. Strain 2 be cultured because it cannot synthesize biotin and methionine. But when both of them are kept on a minimal medium curiously they survive (normal expectation is, they should not live). On nutritional analysis, it was revealed that they were Paratrophs. The formation of paratrophs from two polyauxotrophic varieties can be explained only on the possibility of sexual recombination between strain I and 2.

$B^+M^+L^+ T^+$ $B^-M^-L^-T^-$

1 2

Because of recombination of I and 2 the following combinations may be obtained.

$B^-M^-L^-T^-$ and $B^+M^+L^+T^+$

(a) (b)

Of the above (a) will not survive because it has lost the capacity to synthesize all the nutrients but (b) will survive because it has become a paratroph.

Generic transformation. Every individual born out of Sexual reproduction will have the mixture of traits of both the parents. In genetic terms, this means that there is genetic recombination in the offsping. Genetic recombination or transformation which is an index of genetic change has also been reported in bacteria.

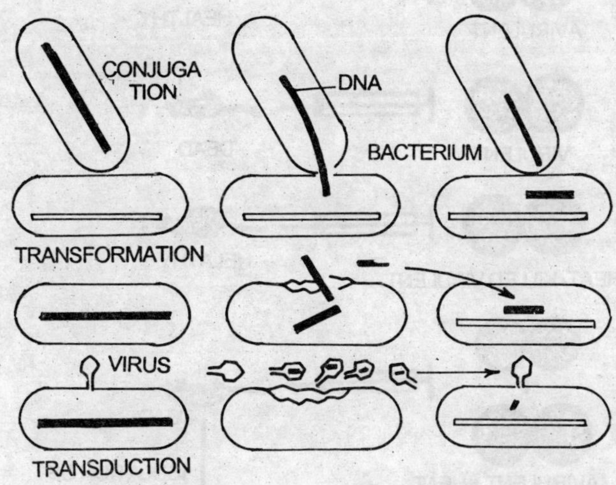

Fig 6.13 Bacteria
Genetic recombinatiom in bacteria

Genetic transformation ie., change of one genetic type to other was first reported by British bacteriologist F.Griffith (1928) while working with pneumococci (Diplococcus pneumoniae).

There are two strains in *Diplococcus penumoniae.* One is capsulated and the other one is non capsulated ones form glistening smooth (S) colonies on the culture medium while the non capsuleatedones form dull, rought(R) colonies. Further, the capsulated (S) strains are virulent while the non capsulated (R) strains are non virulent ie., do not produce toxins. The ability of producing or not producing a capsule is a genetic character.

Griffith injected experimental mice with R strains of pneumococci and found that they survived but whenS strains are injected the, mice died. When heat killed (killing the bacteria by applying heat when DNA gets denatured and will not produce toxin) 'S' strains were injected the mice survived. Griffith next injected both R strain and heat killed 'S' strain to the same experimental mice and they (mice) died. Blood analysis of the dead mice (which were infected withR strain and heat killed S strain) showed the presence of live' S strains

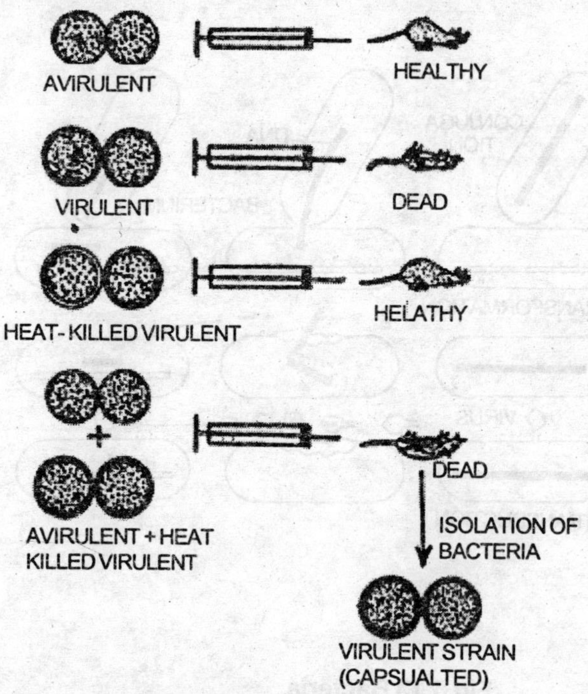

Fig 6.14 Bacteria

The transformation experiment of Gifth with pneumococcus

There are three possibilities how the heat killed 'S' strains can become live again.

I • Heat killed S strain might have become alive in the body of mice- this is an impossibility.

2- A mutation might have occured in the R strain converting them to S strain. This is also not possible because mutations wiB not occur frequently, but everytime the experiment showed the presence of capsulated S strains in the blood of the dead mice.

3. Some factor in the heat killed strain has transferred the genetic formation for capsule synthesis to R strains making them produce the capsule-.If other words they become S strain. This is the only way a live S strain could be found in the blood of mice. Griffith (1928) thought that the transforming principle was bacterial polysaccharide. But Avery et al (1944) demonstrated that the transforming principle was DNA.

This experiment clearly demonstrates that a piece of DNA responsible for capsule synthesis recombined with the DNA of the R strain inducing it to produce a capsule. This can be called a natural method of genetic engineering.

Transduction. Transfer of genetic material from one bacterium to another mediated by bacteriophages is known as transduction. Transduction was first discovered by Zinder and Lederberg (1952) in *Salmonella typhinmurium* and subsequently by other workers in *Escherichia, Shigella, Staphylococcus* etc

. In transduction, when bacteriophages infect a bacterium they often incorporate a portion of the genetic material of bacterium into their genetic system. If such a bacteriophage infects another bacterium (of the same species) the bacterial genes (of the first infection) may get incorporated into these bacteria. These bacteria now exhibit some of the traits of the first strain. This shows that bacteriophates have picked up some genes from the first bacterium and transfered it on to the second bacterium.

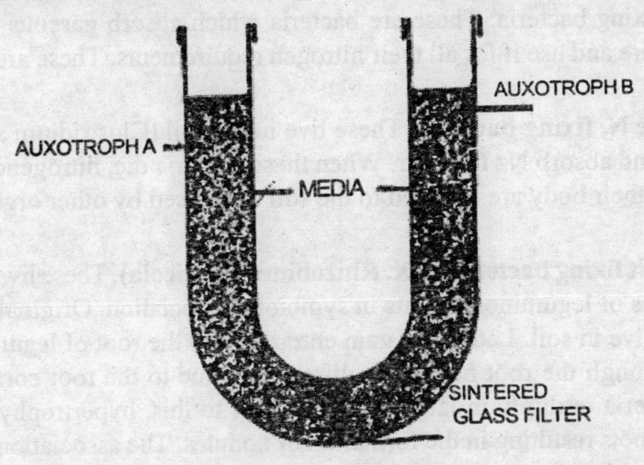

Fig 6.15 Bacteria
The transduction (U tube) experiment of Zinder and Lederberg

In order to prove bacteriophage mediated genetic transfer. Zinder and Lederberg performed a 'U' tube experiment. The experimental set up consists of a 'U' tube where the two arms are separated at the base by means of a sintered glass filter (Bacterial filter). Zinder and Lederberg (1952) grew two different strains of bacteria *Salmonella typhimurium* in the two arms. One strain was infected with bacteriophates. A little later when the bacteria in both the arms of the 'U' tube were analysed it was noticed that the strins in both the arms had become recombinants. Since the bacterial filter does not allow the migration of bacteria, the only way for transfer of genetic material was through bacteriophages since they can move across bacterial filters.

ECONOMIC IMPORTANCE

From the point of view of human welfare, bacteria are doubly important in that they are friends as well as foes of mankind. Hence a discussion can be conveniently classified into beneficial activites and harmful activities. **Beneficial activities:**

7. Disposal of organic matter. Sparophytic bacteria are the most efficient natural scavengers that always maintain the surface of the earth clean. They prevent the accumulation of dead bodies by efficiently decomposing them and returning them to their element stage.

2. Role of bacteria in agriculture. Bacteria help in many ways in enriching the soil nitrogenous substances. While no higher plant can absorb gaseous nitrogen and use it as a source of its nitrogen requirement bacteria can do so. Such bacteria are called "Nitrogen fixing bacteria".

A. Nitrogen fixing bacteria. These are bacteria which absorb gaseous N2 from atmosphere and use if for all their nitrogen requirements. These are of two types:

1. Autotrophic N, fixing bacteria. These live in the soil (Clostridium and Azatobacter) and absorb Nz from air. When these bacteria die, nitrogenous compounds in their body are released to the soil to be used by other organisms.

2. Symbiotic Nt fixing bacteria. (EX. Rhizobium radicicola). These live in the root nodules of leguminous plants in symbiotic association. Originally, these bacteria live in soil. Later they gain entrance into the root of leguminous plants through the root haris and ultimately come to the root cortex where the bacteria multiply in large numbers. Due to this, hypertrophy is formed in the roots resulting in the formation of nodules. The association is symbiotic in that leguminous plants provide food and shelter for bacteria. Inturn bacteria provide the plant with N,. It is for this reason that leguminous plants are rich inproteins. It is for this reason again that leguminous plants are employed in crop rotation (crop rotation is an agricultural practice, in which a legume and a cereal are alternatively grown in the tied. If the same crop is grown year by year, the yield becomes less as the soil becomes deficient in particular nutrients).

B. Nitrifying bacteria. These do not add any extra nitrogen to the soil but convert one of nitrogen compound to the other. For ex. Nitrosomons and Nitrococcus can convert ammonia to nitrate and Nitrobacter convertes nitrate to nitirite. In addition to these, there are ammonifying bacteria which decompose dead bodies and release ammonia.

There are other types of bacteria called "denitrifying bacteria" which act on the nitrogenous compounds of the soil and release gaseous N2 to the atmosphere. All the different forms of bacteria play a very important role in the N, balance of nature by occupying key position in N, cycle.

3. Role of bacteria in industries. Bacteria are of great help in industry. It is bacteria that are responsible for the souring of milk and formation of curds. Bacteria are also employed in the making of cheese from milk. Various types of

bacteria are employed in the manufacture of alcohols and acids. A few of them are:

a. *Lactobacillus acidus* - lactic acid

b. *Acetobacter aceti* - acetic acid

c. *Propionobacterium* - propionic alcohol

d. *Clostridium acetobutylicum* - butyl alcohol .

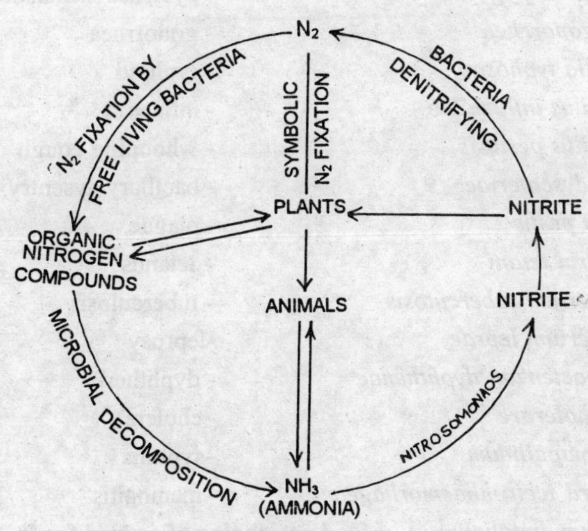

Fig 6.16 Bacteria
The nitrogen cycle

Bacteria are also of great help in curing of coffee, tea and tobacco and in the ning of leather and in the retting of fibres.

The familiar aroma of coffee tea and tobacco will not be there in the raw products. In curing works bacteria eat away the unwanted compounds, producing characteristic flavour. Specific varieties of bacteria are maintained for this purpose and the secret is guarded with utmost care.

In tanneries also, the raw skin is immersed in water containing bacteria. Bacteria eat away the fatty compounds and make the skin soft.

In retting of fibres also, the plant parts with fibres are immersed in water containing bacteria, whereupon they eat away the soft tissues and separate the fibres.

4. Medicinal uses. Bacteria belonging to the order Actinomycetales yield valuable antibiotics.

Streptomycin - Streptomyces griseus

chloromycetin - Streptomyces venezulae

terramycin - Streptomyces rimosus

Harmful activities:

Bacteria can play havoc with human welfare. They cause several deadly diseases. A few examples are given below :

Name of the bacteria	Disease
I .*Streptococcus pygoens*	- pyorrhea (dental decay)
2.*Nisseria gonorrhea*	- gonorrhea
3 .*Salmonella typhosa*	- typhoid
4.*Haemophilus influenZae*	- influenza
5.*Haemophilus pertusis*	- whooping cough
6.*Shigella dysenteriae*	- bacillary dysentry
7.*Pasturella pestis*	- plague
8.*Clostridicrm tetani*	- tetanus
9.*Mycobacterium tuberculosis*	- tuberculosis
9.*Mycobacterium leprae*	-leprosy
11.*Cor~nebacten'um dyphthenae*	- dyphtheria
12.*Vibrio cholerare*	- cholera
13.*Treponemapallidum*	-syphilis
14.*Leptospird icetrohhaemorhageae*	- meningitis
15.*Clostridium botulinum*	- food posionning of canned foods

Bacteria cause disease (to cattle) known as the Anthrax disease (*Bacillus anthracis*).

Denitrifying bacteria bring about loss of fertility of soil by releasing N, to the atmosphere.

Bacterial plant disease:

Bacteria have also not spared the plants. Bacterial plant diseases can be classified as follows.

1. Parenchyma Diseases: These are caused by bacteria which eat away the soft parenchyma tissue of plants. For eg.

A. Angular leaf-spot disease of cotton caused by Xanthomonas malvacearum

b. Citrus canker caused by Pseudomonas citri,

2. Vascular disease or bacteriotracheoses. These are caused by bacteria accumulating in the vascular tissues so that the lumen is clogged and the aerial parts of the plant will die due to lack of nutrient supply for eg:Bacterial wilt of Tobacco caused by Xanthomonas solanacearum.

3. Hypertrophies. Due to the infection of bacteria hypertrophies or galls are formed on the plants. For eg. Agrobacterium tumefaciens causes crown gall disease of apples.

BACTERIAL CLASSIFICATION:

Living organisms are classified into various groups based on resemblances and differences. Taxonomic hierarchies are erected starting with kingdom and ending with species. A species is said to be the lowest stable unit in taxonomic classification. With microorganisms however a species can not have the same rigidity and meaning as in higher organisms. Species are not tight genetic units because very frequently a number of mutations occur in microorganisms and all cannot become new species. Further, extrachromosomal genetic elements initiate chromosonal rearrangements. With all these difficulties however species are still identified and groups are erected. It should be understood however that bacterial groups or orders, families etc are not phylogenetic in the senese they represent discontinuous groups. As genetic changes are quite frequent, phenotypic similarly in many unrelated traits in the basis for categorizing a bacterial species. Bacterial classification is varied in that it is based on several character sources. A few of the important approaches to bacterial classification are described below.

Unit of classification in bacteria:

Like in all other living beings bacteria are also named according to the binomial nomenclature. The first name is the generic name and the second name is the specific name. The following are the units of Classification.

Division

class

order

family

tribe

genus

section

series

species

Three terms are commonly employed in microbiological studies with reference to microorganisms-strain, clone and type species. A strain is a culture of a given species while a clone is a strain derived from a single cell. A type species is a standard one that has been studied in detail and is used for comparison with the unknown. The name of type species usually denotes some special traits of the group.

Artificial classification. This represents the earliest attempt towards classifying living beings* Living beings were classified into plants and animals microorganisms were included under both paint and animal kingdom.

Phylogenetic classification. This is a post Darwinian concept in which organisms are classified according to their phylogenetic relationship. Haeckel, a German biologist placed all the microorganisms under a separate group Protista

as different from plants and animals. R.H. whittaker (1969) developed a phylogenetic five kingdom classification of living beings. Monera, Protista, Fungi, Plantae and Animalia are the five kingdoms identified by Whittaker (1969). Microorganisms are included in the first three kingdoms while all prokaryotes are included under Monera.

Morphological classification. Based on the cell shape, size etc bacteria. Are divided into

a. Higher bacteria - filamentous forms

b. Lower bacteria- unicellular forms

1. Bacillus - rod shaped
2. Spirillum - spiral form
3. Cocci - spherical 4.
4. Comma - comma shaped
5. Cocci can be of the following types

a. *Diplococcus* - cells in groups of two

b. *Tetracoccus* - cells in groups of four

c. *Streptococcus* - cells in chain

d. *Staphylococcus* - cells in sheets

Bacteria may also bse calssified based on various criteria such as staining reaction, nutrition, base composition, nucleic acid hybridization, biostatistical principles etc.

Bergey's manual of bacteriology is the most authentic work on the classification and nomenclature of bacteria.

RICKETTSIAE

The discovery of rickettisiae goes back to 1909, when Howard Taylor Ricketts observed them under the microscope, while examining the blood samples of patients suffering from Rocky mountain spotted fever. These microorganisms were similar to bacteria, but refused to grow on culture media microorganisms from the blood smear of patients suffering from typhus fever, and also from the blood smear of lice fed on typhus patients. In the year 1913, Prowazek, a microbiologist came across similar organisms while studying typhus patients in Siberia. In 1916, De Rocha Lima gave the name *Rickettsia* to the organism responsible for tyhpus fever and he designated the causative organism as *R. prowazekii* in honour of Ricketts and Prowazek who not only discovered the causative organism but also died due to typhus fever during their investigations.

Classification

The classification of Rickettsia is somewhat difficult. As described in the 1957 edition of Bargey's manual of determinative bacteriology, Rickettsiae are included in a separate class

Microtatobiotes includes viruses also, with a single order *Rickettsiales*. The order is divided into four families – *Rickettsiaceae, Chlamydiaceae, Bartonellaceae* and *Anaplasmataceae.*The family Rickettsiaceae is fairly representative of the order and includes two genera *Rickettsia* and *Coxiella.*The genus *Rickettsia* has a number of species causing typhus, trench fever, spotted fever etc. while *Coxiella* causes the fever. The following table summarizes the causative organisms, disease, host, vector etc.

Distinguishing Characters

Rickettsiae are obligate parasite like viruses. They are also intracellular. *R.qunitana* is an exception in that it grows extra cellular in the louse gut and

Table : Summary of Pathogenic organism in various rickettsioses, their distribution hosts and vectors

Category	Organism	Disease	Geographical Distribution	Arthropod Vector	Vertebrate host
Typhus group	*R. prowazekii*	Edidemic typhus and	Worldwide	Body louse	Man
		Brill-Zinsser disease (recrudescent typhus)		–	Man
	R. mooseri (typhi)	Murine or endemic	Worldwide	Rat flea	Rats
Spotted fever group	*R. rickettsii*	Rocky Mountain spotted fever	Western hemisphere	Tick	Small wild mammals, dogs, birds
	R. sibirica	Siberian tick typhus	Siberia, Mongolia	Tick	Wild animals and birds
	R. conori	Boutonneuse Fever Small wild mammals,	Mediterranean littoral	Tick	dogs
	(R. parkeri)	–			
	R. akari	Rickettsial pox, vesicular rickettsiosis	North-east U.S.A U.S.S.R.	Gamasid mite	House mouse
	R. australis	Queensland tick typhus	Australia	Tick	Bush Roents
Scrub typhus	*R. tsutsugamushi*	Scrub typhus	Oceania	Trombiculid mite	Small wild rodents and bords

Category	Organism	Disease	Geographical distribution	Arthropod Vector	Vertebrate host
Trench fever	*R. quintana*	Trench fever	Europe, North Africa, Middle East, Mexico	Body louse	Man
Q fever	*Cox, bumetii*	Q fever	Worldwide	Two cycles : (1) Tick and small wild mammals (2) Cattle, sheep and goats by aerosol and milk without atrhropod-vector	

has been cultivated on bloodsugar. Others can be cultivated only in live tissues (tissue culture medium). Rickettsiae are like bacteria in many respects, though they are obligate parasite like viruses. Electron microscopic studies reveal that rickettsiae have the same structural features like bacteria.

The following are some of the salient features of rickettsiae.

1. They are minute in size (smaller than most bacteria but still visible under light microscope), and have a characteristic staining capacity.
2. The cells are plemorphic and may be rod shaped or occasionally filamentous.
3. The size varies – diameter from 200nm to 500nm and length 800nm to 2.0 μm.
4. They are mostly gram negative, except *Coxicella brunetii* which is gram positive.
5. Cell wall has muramic acid.
6. DNA, RNA and many enzymes are present in the cell. The ratio of DNA to RNA is 1 : 3.5.
7. Reproduction is by binary fission.
8. Rickettsiae includes both filterable (*Coxiella*), and non filterable (*Rickettsia*) organisms.
9. It is sensitive to many antibiotics including chloromycetin.
10. Rickettsiae cannot multiply outside the host cell. They can be cultured on the yolk sac of chick embryo.
11. Rickettisae multiply in the cytoplasm of host cell except spotted fever group which multilpy in the nuclei also.
12. They produce only endotoxins – no exotoxins.

Similarities and differences between Rickettsiae, Mycoplasma and Viruses

1. Rickettsiae and mycoplasma can be stained and identified under a light microscope, whereas viruses can be studied only under election microscope.
2. Rickettsiae have rigid cell wall while mycoplasmas do not have a wall.
3. Rickettsiae and mycoplasma have their own enzyme machinery unlike viruses.

Morphology and biological properties

In their morphological appearance, rickettsiae are like small bacilli or cocci. Electron micrographs also confirm their bacterial nature. Rickettsiae cultivated in yolk sac of chick embryos are either rod like or rounded ranging in size from 0.5μm. The cell wall is the outermost and is double or multilayered. This surrounds a hyaline envelope or membrane enclosing a dense cell body. The cell wall is about 7–10 nm thick and the cytoplasmic membrane is 6–8 nm in thickenss, and has the typical unit membrane structure seen in the bacterial cell. Some rickettsiae like *R.prowazeki,* have an ether soluble capsule. Internal to the cell membrane, there are ribosomes (7-20) nm thick which are presumably of the same type found in bacteria.

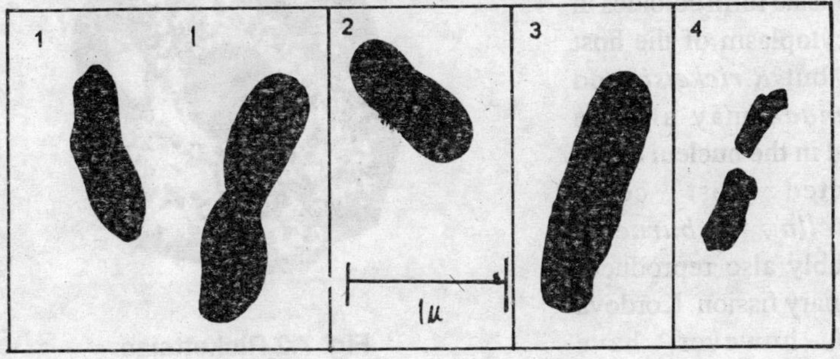

Fig. 7.1 Rickettsiae

1. Rickettsiae of epidemic typhus
2. Rickettsiae of endemic typhus
3. Rickettsiae of Rocky Mountain spotted fever,
4. Rickettsiae of American Q fever

The nuclear material may be organized in a central body with radiating fibres (*Coxiella burnetii*), or dispersed as a reticulum of fibrillar structures. Double stranded DNA is found in *C. burnetti.*

Chemically, the cell wall of rickettsiae consists of glucosamine and muramic acid. In addition, diamnopimelic (found in gram negative bacteria) is also found in the wall. This points to a close relationship between rickettsiae and gram negative bacteria.

Rickettsiae (except for *C.burnetii*), are unstable outside the host cells and their cytoplasmic membranes readily leak macromolecules such as RNA, and intracellular ions into outside aqueous media. Suspensions of rickettsiae outside of host cells have a variety of enzymes involved in metabolic pathways connected, to respiratory break down of carbohydrates. Metabolic pathways for the synthesis of lipds and proteins are also present in rickettsiae. Moulder (1962), has worked in detail on the metabolic basis of intracellular parasitism in rickettsiae.

Physiological studies of Ormsbee (1969), suggest that the host cell contribution to the rickettsiae include substrates such as glutamate, pyruvate, ATP, NAD and Coenzyme A.

Reproduction

The Rickettsiae multiply intracellularly by means of binary fission except *R.quintana*. Most rickettsiae form colonies in the cytoplasm of the host cell, but *R.ricketsii* and *R.canada* may also be found in the nucleui of the infected host cells. *Coxiella burnetii*, probably also reproduces by binary fission. Kordova *et-al*, however, have suggested a different mode of division involving smaller filterable forms.

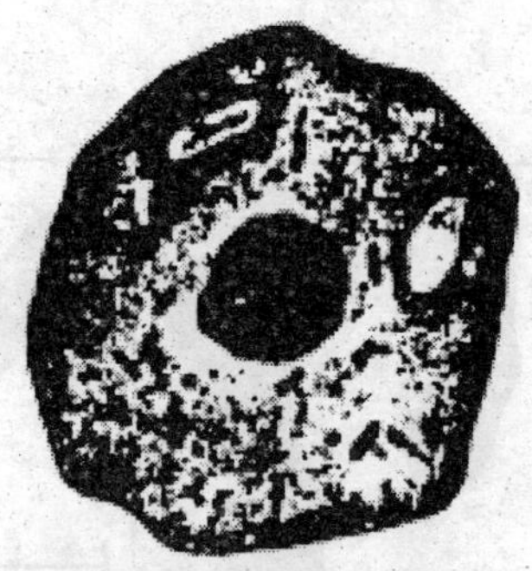

Fig. 7.2 Rickettsiae

Some criteria used in the identification and classification of bacteria.

Typical rickettsiae are indicated by arrows

Staining Characters

Rickettsiae generally stain feebly with the usual dyes, and they are gram negative. Special stains like Giemsa or Macchiavello used for blood smears of protozoa can be used to demonstrate the pathogen in the infected tissues.

Laboratory Culture

Rickettsiae cannot be cultured in cell free medium. They can, however, be cultivated in agar tissue cultures where they develop within the growing cells of the culture just the same ways as they do in *vivo*. They can also be cultured on the chorioallantoic membrane of the chicken embryo (same procedure as used for cultivation of viruses.) The most prevalent and successful method, however, is to grow it in the yolk sac of the growing chick.

Rickettsiae have also been multiplied in mouse fibroblasts, Detroit 6 cells, Hep2, Hela and other continuous cell lines. Of all the rickettsiae, *R.quintana*, is an exception because it can grow extracellularly on blood agar.

Metabolic activity and Antibiotic sensitivity

Pure cultures of rickettsiae may be obtained by high speed centrifugation of infected yolk sacs. The organisms show the ability of oxidation of pyruvic acid like bacteria, but unlike viruses.

Among the antibiotics, *sulfonamide drugs are not to be used as they stimulate the growth of rickettsiae instead of inhibiting it.* Para amino benzoic acid (PABA), is known to inhibit their growth. Chloramphenicol and Tetraclines are very effective against rickettsiae. They probably interfere with some essential enzymic activity.

ECONOMIC IMPORTANCE

Rickettsiae cause five types of infections in human beings. These are – Typhus group fever, Spotted group fever, Scrub typhus, Trench fever and Q fever. Brief descriptions of these is given below.

Typhus fever group

There are three kinds of fever in this group. These are – Epidemic typhus, Brill Zinsser disease and Endemic typhus.

Epidemic typhus

This is caused by *Rickettsiae prowazekii* and has been reported from all parts of the world including India (Kashmir). Man is the only natural vertebrate host. Human body lice (*Pediculus humanus corporis*) is the vector. Lice gets the infection by feeding on typhus patients. Rickettsiae multiply in the gut and appear in faeces in about 4–5days. When a healthy man contacts the infected lice or even inhales the faeces of lies, infection is introduced. After an incubation period of 5–21 days fever and chill begin followed by rash. The rash starts on the trunk and spreads over limbs. Patients become delirious and stuporous. If untreated, fatality is 15 to 70%.

Brill Zinsser

This is a mild typhus like disease, found in Jewish migrants of New York (USA).

Endemic typhus

This is a milder form of epidemic typhus caused by *Rickettsiae Mooseri.* The infection is transmitted from rat to man by fleas. Fleas remain unaffected but transmit the infection. Man acquires the infection through the bite of infected flea or through consumption of food contaminated by rat's urine or flea faeces etc. There is no spread of infection from man to man.

Spotted fever

There are several fevers in this group (see table on page 265), but the Rocky mountain spotted fever is the most serious type. Clinically, it resembles the epidemic typhus. Lesions spread all over the body including the palm, sole and buccal mucosa. *R. rickettsii* is the causative agent.

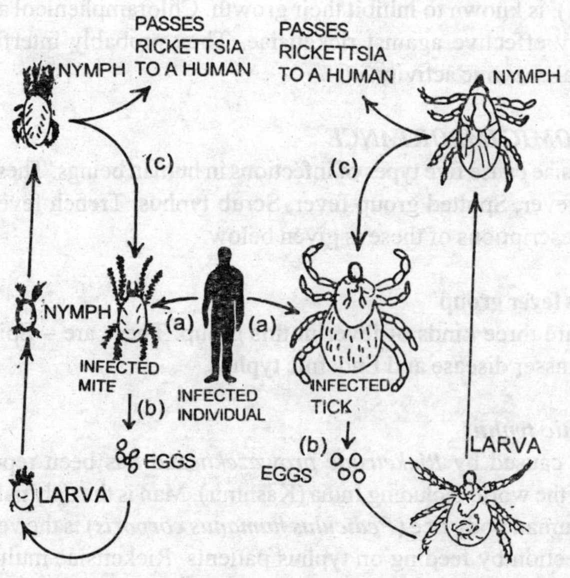

Fig. 7.3 Rickettsiae
Maintenance of rickettsial diseases in nature by ticks or mites.
(A) Tick absorbs the blood of infected man and acquires rickettsiae,
(B) It passes the organisms to its eggs
(C) Eggs develop into infected larvae, nymph, and adult stages, after which the cycle repeats itself. The mite or tick my pass the infection to another human being.

Scrub typhus : It is caused by *Rickettsia tsutsugamushi (tsutsuga* = dangerous; *mushi* = insect or mite). The disease was first observed in Japan. The disease is transmitted by the vector mite *Trombicula akamushi* in Japan, and by *T.deliensis* in India.

Human beings are bitten by mite larvae. The mite feeds on the serum of warm blooded animals only once during its life-cycle, while the adult mite feeds on the plant. Rickettsiae are transmitted transovarially in mites. Rodents acas reserviors. When a mite bites the rodents, and in turn the human beings, the pathogens are transferred to the human system. Eschar develops at the site of bite in human beings. After transfer to the human system, rickettsiae take about 1–3 weeks for incubation and multiplication. Then there is high fever, headache and conjuctivital infection. If untreated, fatality rate is 0.5 – 60%.

Trench fever

This is an exclusive human disease. No animal reservoir is known. The disease is transmitted by body lice. Faecal matter of lice carries the pathogen in 5–10 days after contracting infection. Lice is unharmed, but it remains a carrier of infection. The disease is caused by *R. qintan* (quintana a five-day fever) and leads to chronic *rickettsiemia.*

Q Fever

This is caused by *Coxiella burnetii,* and seems to be world wide in distribution. Bandicoot is the reservoir, and the vector is Ixodid tick. Transovarial infections have been transmitted by ticks, from where it reaches the cattle, sheep and poultry. Human infection is acquired by inhaling the dust contaminated by rickettsiae, derived from animals and their products. Infection may even occur by consumption of infected milk. Interstital pneumonia charactarizes the disease. Rashes on the skin are not observed. Liver, heart and other organs may get infected. Recovery is spontaneous. Man to man transmission is very rare.

Treatment of disease caused by Rickettsiae

Chloramphenicol and tetracyline are very effective, but the treatment should begin quite early after the detection of the disease. The antibiotics do not remove reckeusiac but suppress there growth. The immune system of the patient plays an important role. in the control of the disease. Active immunization is possible by using vaccines for epidemic typhus, spotted fever and Q fever.

Ecology and geographic distribution of Rickettsiae :

The geographical distribution of rickettsiae has been given in table (see page 265). Q fever practically has a world wide distribution coinciding with the distribution and movement of domesticated animals. The wide distribution of rats and their ectoparasites (fleas, ticks, tesete) determines the prevalence of typhus which seems to be limited to SE Asian subcontinent and Australia The distribution of arthroped hosts determine the prevalence of the spotted fever group. Epidemic typhus distribution depends upon the social conditions, poverty etc. of the area as it is directly linked to the head lice which thrive on personal unhygenic conditions, overcrowding etc.

In United Kingdom, Q fever is prevalent. This is more so in places where infected cattle and human beings stay together.

Several Ecosystems are responsible for the maintenance and distribution of various groups of rickettsiae. The interaction of these members of the ecosystem with the social patterns of human beings determine the prevalence of the disease.

Epidemic typhus is a result of association of lice, which is directly related to unhygenic conditions. In murine typhus, the inter action includes – rat, flea and man. In scrub typhus, the components involved are small wild animals or birds, tick and man. In Q fever again small wild animals, ticks, cattle or sheep and man are involved. A comprehensive study of all these hygeine is necessary if one wants to excerise greater control on the spread of the disease.

Evolution of Rickettsiae

A general survey of rickettsiae reveal that they are habitually residents in the body of various insects and other arthropods. It is widely believed that – rickettsiae were originally parasites of arthropods only, and have only become secondarily parasitic on human beings and other mammals, as a consequence of blood sucking habit of mites, ticks fleas etc. Now these arthropods are functioning as vectors, transmitting rickettsiae into animals and man. Many non pathogenic rickettsiae have also been reported to be harboured by various arthropods.

CYANOBACTERIA

(Blue Green Algae)

The cyanobacteria are a group of photosynthetic prokaryotic organisms that have been traditionally classified under algae as a separate class cyanophyceae (myxophyceae). But the recognition of the fact that all blue green algae are prokaryotic while other members of algae are eukaryotic has made the inclusion of blue greens among algae untenable. In the modern five kingdom system of classification, blue greens are included in the kingdom Monera along with true bacteria. (For the purpose of easy understanding however, throughout this chapter the word blue green algae is used.

The class includes the most primitive members among algae which have the following general characters :

(1) Members are of the blue green color due to specific pigmentation.
(2) Absence of definite plastids.
(3) Absence of motile phase.
(4) Absence of sexual reproduction.
(5) Nuclear material being diffused in the centre of the cell forming an incipient nucleus.

The class comprises about 165 genera and nearly 1500 or more species.

These members are commonly called blue green algae because the cell contains the bluish green pigment phycocyanin, and red pigment phycoerythrin, in addition to chlorophyll, myxoxanthin, myxoxanthophyll and carotenoids etc.

Occurrence and Distribution

One of the unique features of blue green algae is their capacity to withstand the vagaries of the environment. They possess a wide range of tolerance to the climatic fluctuations, and as such they are found inhabiting almost all types of ecological niche of the environment.

A good majority of these occur in a variety of fresh water habitats like hard water lakes, streams and wet soil, rocks, damp barks of trees etc. In fresh water forms, members may be purely aquatic or semi-terrestrial. *Nostoc, Anabena, Gloetrichia, Rivularia* etc. are usually found fresh water ponds and ditches.

There are some members which are marine,–ex. *Dermocarpa, Trichodesmium,* etc., where they form an important constituent of planktonic flora (floating plants).
Some of the planktonic flora occur in large numbers in water so as to form "waterblooms"(The excessive growth of these algae is referred to as "waterblooms"). In marine forms also, the growth may reach such proportions as to impart a characteristic coloration to water (ex. Red sea is called so because of the presence of *Trichodesimium erytharacum).*

There are subterranean forms also growing at a depth of 50 cms, sometimes as low as two meters (Moor and Carter, 1926). They can also occur at temperatures as high as 85c which other plants cannot withstand (Setchell 1926). These are known as thermal algae. At the bottom, the salts–calcium and magnesium are precipitated and these are known as *"travertine"* which gives a variety of colors eg., *Schizothrix.*

Some of the blue greens like *Nostoc, Anabena* can be symbiotic in association with higher plants like Bryophytes *(Anthoceros),* and Gymnosperms *(Cycas),* in addition to their independent existence. *Nostoc* is found in the mucilagenous cavities of the thalli of *Anthoceros and Notothylas.* It is also found in the sporophylls of *isoetes* (Sundara Rajan, unpublished). *Anabena azollae* is found in *Azolla,* and *A. cycadecae* is found in the coralloid roots of *Cycas.* In all these cases the relationship is supposed to be symbiotic. While the blue green algae get shelter and protection, the higher plants are benefitted by the N, fixing capacity of these algae.

Many blue green algae are also found as phycobionts (Algal partners), in the lichen association. The lichen thallus consists of an alga and a fungus. The relation between the two partners here is known as Helotism, a degree of symbiosis. Some blue greens may also be found as parasites (40 sps. of Oscillatoriaceae), in the alimentary canal of man and. other animals.

Thallus construction : Among the algae the blue greens are the lowest and possess the simplest structure. The plant body may be classified into 3 types, namely :

(1) Unicellular forms
(2) Non-filamentous colonial forms and
(3) Filamentous forms.

(1) Unicellular forms : Members of chroococcales are mostly unicellular, ex. *Chroococus, Tetrapedia* etc.

(2) Non filamentous colonial forms : In many members, the cells fail to separate after cell division, as a result the cells remain together forming colonies. These are loosely formed colonies and the cells are held together by the gelatinous matrix. Thus, the colonies are of various shapes like irregular, spherical, squarish etc. The shape of the colony depends on the plane of division and the effect of the environment, which determines the adhering capacity of the mucilagenous sheath. Ex., *Merismopaedia.*

(3) Filamentous forms : The continuous division of a single cell in only one plane results in the formation of a filament, the row of cells in the filament is called a "trichome". A "trichome" may be of the same diameter throughout *(Oscillatoria),* or may taper towards the tip as in *Rivullaria.* Trichomes may be branched, *(Oscillatoria),* unbranched *(Nostoc),* or may show false branching as in *Scytonema.*

Locomotion

Only a few genera exhibit movements. Unicellular and filamentous forms show locomotion. In the former, the movement is slow and later in the rapid. Among the filamentous forms which exhibit locomotion are *Oscillatoria, Trichodesmium, Sporulina, Arthrospira, Micrococcus, Anabena* (Castle 1926), and *Cylindrospermum.* (Geitler 1932). Among the unicellular forms *Syneclwcoccus* (Perfilier 1915), and *Chroococcus turgidus* (Lund 1950) are important.

The movements may be of creeping or gliding types, in the longitudinal axis backwards and forwards. Various theories have been put forward to account for these movements

1. *Secretiontheory(Gaiduko and Scbimid,* 1910,1918): The protoplasmic streaming of cell contents is seen in the cells which is 'related to the movement through the secretion of mucilage.

2. Many algalogists think that movement is brought about by the rythmic longitudinal waves caused by alternate contraction and

expansion traversing the entire length of the trichome (Phillips, 1904).

3. Hansgrig (1887), attributed the movement to diurnal changes resulting from a turgor gradient all through the trichome.

Cell structure

In a great majority of the blue green algae, two regions are distinguishable within the protoplast. There is a peripheral region (chromatoplasm), within which lies the photosynthetic pigments and a colourless central region (central body, centroplasm or incipient nucleus). But this distinction may not always be obvious. As in *Chroococcus turgidus,* the pigments are distributed throughout the protoplast, this, however, is an unusual condition.

Each cell is externally bounded by a cell wall made up of cellulose and pectic compounds. This is surrounded by a gelatinous sheath of varying thickness. The sheath also is made up of pectic compounds, and may be thick and firm or may be thin and inconspicious. The sheath shows stratification which may be due to differential hydrolysis in the sheath. It may be colourless or variously coloured. The sheath is supposed to perform the following functions:

1. Increase the water retaining and water absorbing capacity so that they can thrive in dry habitats also.

2. Coloured sheath may protect the cell from the injurious effect of intense.

Chromatoplasm : The cytoplasm may be alveolar (Baumgartel, 1920), or homogeneous, (Geitler, 1936). The outer region which is called chromatoplasm is coloured and consists of various pigments like Chlorophyll, Carotin, Myxoxanthin, Myxoxanthophyll, Flavisin phycocyanin, Xanthophyll and Phycoyerythrin. Also present in the Chromatoplasm are oil drops, glycogen and few spherical bodies known as cyanophycin granules which are proteinaceous in nature.

Philips (1904), Olive (1904), and others think that blue green algae possess a peculiar type of Chromatophore. This is supported by electron microscopic studies. The Chromatoplasm has a lamellated structure which is thickened; these granular thickenings are considered similar to the grana of higher plants.

Complementary Chromatic adaptation (Gaidukov phenomenon) : Some blue green algae have a capacity to change their colour in relation to the wave length of the incident light. This is known as complementary chromatic adaptation. This is of great use to them in using all ranges of light.

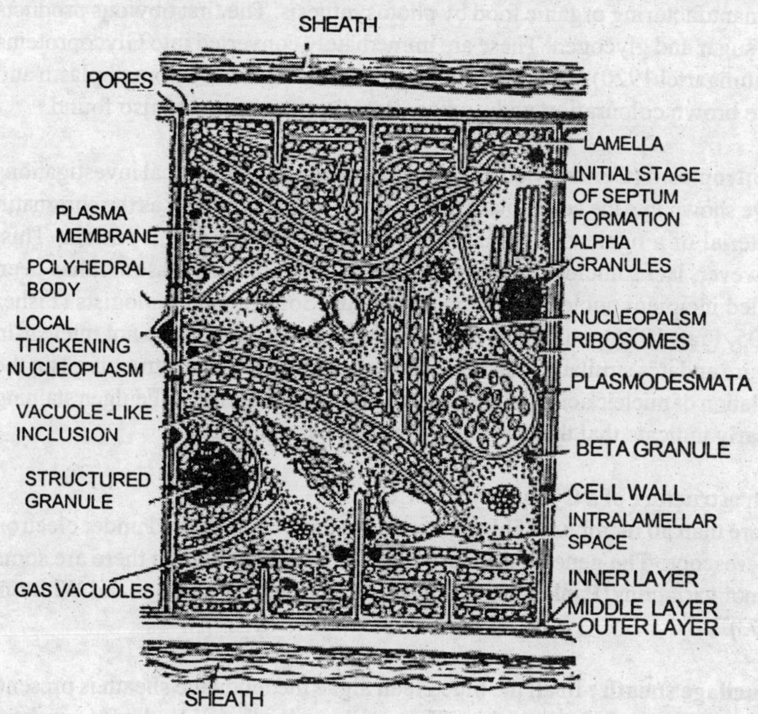

Fig. 8.1 Cyanobacteria
Electron microscopic structure of cyanophycean cell

Gas Vacuoles or pseadovacuolcs : These are often present in the chromatoplasm as in *Policy stis* and *Anabena.* The presence of these, imparts a yellowish green color. They appear as black bodies under low power, and red in high power view of microscope probably due to changes in refraction.

According to Strodmann (1895), these are really gas filled cavities, the gas might be nitrogen or a nitrogenous compound. Klenmann suggests that the vacuole membrane is permeable to oxygen and the gas oxygen is helping in respiration.

Photosynthesis : Since chloroplast is present, blue green algae are capable of manufacturing organic food by photosynthesis. The first obvious products are sugar and glycogen. These are immediately converted into Glycoproteins (Baumgartel 1920). These products can be located in the chromatoplasm and give brown colouration with Iodine. Besides these, fats are also found.

Centroplasm (Central body, incipient nucleus) : Cytological investigations have shown that the centroplasm stains much in the same way as the chromatin material of a nucleus. This is composed of DNA, RNA and protein. This, however, lacks nuclear membrane and nucleoli. For this reason, it has been called incipient nucleus or an open nucleus. Some algal cytologists (Fisher 1905, Pratt 1925), adhere to the view that the central body is not nuclear in nature and it is similar to chromatoplasm in its fundamental structure. But the isolation of nucleic acids (Mockbridge 1927), and the positive Feulgen staining clearly indicate that the centroplasm is nuclear in nature.

Ultrastructure of a Cyanophycean Cell

More than 50 members of blue green algae have been studied under electron microscope. The general structure is as follows even though there are some minor variations (Pankratz and Bowen 1963; Fogg, 1972; Can- and Whitton, 1973).

Mucilage sheath : In all the blue green algae the mucilage sheath is present, forming the outermost boundary. The mucilage sheath may be thin *(Anacy'stis)* or thick *(Anabena)* and is made up of many cellulose mirofibrils arranged randomly in a homogeneous matrix, According to Fogg (1973), the fibrils are made of pectic acids and mucopolysaccharides. Sugars may also occur in the mucilage sheath *(Nostoc muscorum)*.

Cell Wall : This is rigid and is made up of two layers. The outer wall layer is usually folded while the inner is smooth. The inner layer is further distinguishable into two layers. The inner one is made up of mucopeptides and is digested by Lysozymes, a feature common with the bacterial cell wall. According to Jost (1965), Can- and Whitten (1973), the cell wall is a rigid complex made up of four layers viz., L , L , L, and L - The layer L is an eletron transparent zone.

The plasma membrane found internal to the cell wall is made up of two layers, separated by a less opaque zone.

Photosynthetic apparatus : The peripheral region of cytoplasm (chromatopfasm) consists of thylakoids (lamellae) of uniform breadth arranged in two or more longitudinally arranged stacks. Each lamella is 70–80 A^0thick and is a double membranous structure.

In *Anabenopsiscircutaris,* the thylakoid lamellae traverse the entire cytoplasm, including the centroplasm in the form of a network. Some algalogists (Bisalputra et a,l 1969), opine that the lamellae are also seats of cellular respiration and as such they are called photosynthetic respiratory membranes. Particles of phycobilisome and biliproteins are found attached to the surface of the thylakoids (Grant, 1975; Godwin, 1974).
Dicotysomes, mitochondria and endoplasmic reticulum are absent in cyanophycean cells.

Other Cell inclusions : Gas vacuoles, ribosomes, and many other polyglucan granules are found in the cytoplasm. Recent studies have indicated that the gas vacuoles are irregular in shape and are made up of cylindrical vesicles with conical ends. These vesicles are stacked in bundles and are enveloped by a single membrane.

Walsby and Foggy (1975), have shown that metabolic gases are present in these vacuoles and they regulate the buoyancy of plankatonic members *(Oscillatoria, Anabena* etc).

Ribosomes present in the cell take part in protein synthesis.

Centroplasm (Nucleoplasm) : The centre of the cell is occupied by a : three dimensional network of the genetic material–DNA. This region is somewhat transparent and consists of numerous DNA microfibres arranged randomly. Histones and protamines are not present. There is no nucleolus. Each DNA fibril is– $2A^0$ in diametre and is similar to what is found in bacteria and other dinoflagellates.

Other granules : There are other granular bodies present in the cytoplasm besides the ones mentioned above. These are called Cyanophycean granules, polyhedral bodies etc. The polyhedral bodies are glycogenic in content (Lang, 1972), and associated with thylakoids, while the cyanophycean granules represent sites of reserve food material.

Reproduction : While generally one can say that there is no sexual reproduction and no gametes are formed, there are, however, specific instances of genetic recombination indicating a type of sexual reproduction much the same way as in bacteria. Genetic recombination has been demonstrated in *Anacystis nidulans* (Kumar H.D. 1962), and *Cylindrospermum majus* and other filamentous forms (R. N. Singh & R.N. Sinha 1965).

The most prevalent methods of reproduction are vegetative propogation and formation of nonmotile spores.

Vegetative Propagation

(1) Cell Division : The cell division starts with the formation of a ring like outgrowth which grows inwards dividing the cell into two. Simultaneously, the chromatin material also divides and is distributed to the two daughter cells. There is no formation of chromosomes or spindles.

(2) Hormogones : This is formed in the filamentous species. Hormogones are usually bitsoftrichomes consisting of a few cells. Hormogones are separated from the trichom~by the formation of bi-concave separation discs. These are very common in Oscillatoriaceae. Each hormogone develops into a new filament.

(3) Hormosporcs or Hormocysts : At the time of formation of a Hormocyst a large amount of food is accumulated in the hormogones and they become thick walled. These are usually formed at the tips of the trichomes. They germinate to form a new filament.

Spore Formation

(1) Akinetes, Resting spores or Arthrospores : These are nonmotile spores generally formed in Nostocaceae and Rivulariaceae, to tide over the adverse environmental conditions. At the time of formation of an akinete, the cell accumulates large amount of food and the cell wall becomes very thick. The wall may be distinguished into outer endospore and inner endospore. This type of spore in which the original cell wall forms the outer most layer is known as " Akinete". Akinetes may be isolated in a trichome or may be formed in a series. These are extremely resistant to environmental hazards and remain viable for long periods. On germination, an akinete develops into a new individual.

(2) Endospores : These are also called gonidia and are produced in all the genera of chamaesiphonales and in a few members of other orders. "These are usually spherical but may be angular also. They become free when the cell wall decays. Unlike akinetes, endospores germinate immediately.

(3) Exospores : In the epiphytic member *Chameosipnon,* exospores are generally formed. After a certain period of growth the protoplast abstricts externaly cutting of a spore much the same was as a conidium is formed in fungi.

(4) Nannospores : In certain forms like *Gloeocapsa,* where the cell division is very rapid the daughter cells are small and are called Nanocysts or Nannospores. They are somewhat similar to endospores. Eventually Nannospores develop into new individuals.

(5) Heterocysts : All filamentous Cyanophyceae except Oscillatoriaceae have special thick walled cells called Heterocysts. They are formed from the vegetative cells. They may occur singly or in chains. In some genera like *Rivularia,* they are always basal while in others \ike *Nostoc,* they are intercalary.

Each heterocyst consists of a thick two layered wall. At the poles, the wall swells to form a nodule known as the polar nodule. Depending on the position there may be one or two nodules; one in basal and two in intercalary. As the heterocyst matures, the protoplast inside becomes more and more transparent due to its transformation into a homogenous viscous substance.

Various views have been put forward to account for the nature of heterocysts. One view is that the heterocyst is spore like. Brand (1933), Geitler (1911) Canabouse (1929), and Desikachary (1946), have reported exceptional cases where heterocysts have germinated to give rise to new individuals *(Nostoc, Anabena* etc.). Spratt (1911) has shown heterocysts forming a number of endospores as in *Anabena cycadicae.* These show that heterocysts are archaic reproductive structures which are largely functionless now.

Although functionless, heterocysts have taken a number of secondary functions. These are:

(1) Helping in the development of Akinetes (Fristch, 1951).
(2) Help in the breaking up of a trichome (Kohl, 1903).
(3) In genera with false or true branching, a relation exists between the origin of the branches and the position of the heterocyst.
(4) Help in N_2 fixation.

Genetic recombination

It is generally believed that sexual reproduction is absent in blue green algae, but, genetic recombination 'has been reported in several members. According to Kumar (1962), biochemical recombination seen in the blue green algae is a kind of parasexual phenomenon, and it differs from true sexuality in that "it is not attended by syngamy ormeiosis", but the true function of sexuality viz., gene recombination is achieved. Gene recombination seen in blue green algae is similar to what is seen in bacteria.

Genetic recombination was first demonstrated in *Anacystis nidulans*, (Kumar, 1964) *Cylindrospermum majus* (Singh and Sinha, 1965), *Anabena dolislum* etc. Genetic recombination has been indicated with reference to antibiotic resistance and such other markers. Kumar (1962) cultured strains resistant to streptomycin and penicillin on separate media, and obtained a resistant variety as shown below:

Strain 1- S^+P
Strain 2- S - P^+

Strain 1 is sensitive to penicillin, but resistant to streptomycin. Strain 2 is sensitive to streptomycin but resistant to penicillin. When these strains were a medium containing both streptomycin and penicillin, surprisingly some cells survived and they possessed resistance to both the antibiotics (s^+p^+). This is possible only by the recombination of strain 1 and strain 2. Pikalek (1962), raised an objection to the experiment mentioned above by stating that penicillin used by Kumar (11964), was not a stable one.

Brazin (1968), was able to demonstrate unequivocally gene recombination in *Anacystis nidulans* using single resistant strains with reference to polymixin B and streptomycin. He could isolate double resistant strains in both polymixin B and streptomycin.

Nothing definite is known about the mechanism of genetic system involved in gene recombination. According to Kumar (1982), it is likely that recombination is brought about by conjugation between donor and recipient cells much the same way as in bacteria.

Lazaroff and Vishniac (1962), have claimed that under certain conditions, filaments of *Nostoc muscorum* actually fuse, much in the same way as heterokaryotics do in fungi.

Cyanophages : Viruses attacking and bringing the lysis of blue green algal cells are called cyanophages.

Kraus (1961) was the first one to hypothesize about viral attacks on blue-green algal blooms causing the disappearance of the blooms.

Cyanophages were first discovered by Safferman and Moris in 1963, in *Lyngbya, Phormidium* and *Plectonema.* The virus was named LPP-1. Since then, cyanophages have been discovered attacking a number of genera like *Anacystis, Chroococcus, Microcystis, Anabena, Nostoc, Cylindrospermum* etc.

Cyanophages are broadly categorised into six viz.,

(i) LPP-1
(ii) LPP-2, and
(iii) SM-I, LPP-2 viruses are similar to LPP-1 except that they are serologically different. SM-I viruses are polyhedral and have shorter tails than LPP viruses.
(iv) N-l virus infecting *Nostoc,* has a contractile tail,
(v) AS-1 infecting *Anacystis and Synechococcus* is the largest phycophage discovered so far, and
(vi) Miscellaneousphycophages infecting genera like *Cylindrospermum.*

In their basic morphology, cyanophages are similar to bacteriophages with a distinct polyhedral head and a short tail. Double stranded DNA is the genetic material. The replication of the cyanophages is almost similar to that of the bacteriophages.

Parallelism between blue green algae and bacteria

Cohn (1853), first recognised the affinities between bluegreen algae (Schizophyceae), and bacteria (Schizomycetes), and placed them under a single division Schizophyta. Later, Copeland (1936), classified the bluegreen algae and bacteria under Monera. For quite a long time however, algalogists like Fritsch opposed this view. But it cannot be denied that blue green algae and bacteria share several characters in their cell structure, reproduction, cytology, chemistry etc. In addition to this, fossils of precambrian rocks indicate that the two groups came into existence almost at the same time. Christensen (1962) introduced the kingdom Procpuyota to include blue greens and bacteria.

The following are the similarities between blue-green algae and bacteria :

1. Nucleus has no membrane and not well orgnanised.
2. Thylakoids are present, but these are not membrane bound as in eukaryotes.

3. Plastids are absent.
4. There is no gamete formation, fusion or meiosis.
5. Genetic recombination is reported in both.
6. Di-aminopimilic acid is found in the cell wall.
7. Mucilage sheath is present in both (Gelatinous capsule is found in some bacteria).
8. Some blue green algae and some bacteria can fix atmospheric nitrogen.
9. N fixing organisms may both be free living or enter into symbiotic association with other higher plants.
10. Symbiotic association with higher plants is called phycorrhiza (algae), and bacteriorrhiza (bacteria).
11. Flagellated phase absent in the life-cycle .
12. They grow easily in sulphurated atmosphere.
13. Transformation and transduction are reported in both the groups.
14. Endospore formation is seen in both taxa.
15. Asexual reproduction mainly by cell division.
16. Cell organelles like endoplasmic reticulum, golgi complex and membrane bound vacuoles are lacking.
17. False branching is seen in some blue green algae *(Syctonema)*, and some bacteria *(Cladothrix)*.
18. Both groups are attacked by viruses which bring lysis of the cell. These are the bacteriophages and cyanophages.

Differences between Bacteria and Blue-green algae

	Bacteria	**Blue-green algae**
1.	Few are autotrophic while a large majority are heterotropic	Photosynthetic
2.	Photosynthetic pigment is Bacteriochlorophyll	Chlorophyll *a*
3.	O_2 is not evolved in phothosynthesis as water is not H_2 donor Water is H_2 donor	O_2 is evolved during photosynthesis.
4.	Unsaturated fatty acids of certain categories absent	Present
5.	No Heterocyst	Heterocyst present in many members

Are blue-greens algae?

The list of similarities enumerated above and the possibility that both blue-greens and bacteria evolved simultaneously have made many people to regard that blue greens are closer to bacteria than to algae, and the blue greens should be placed along with photosynthetic bacteria. As has been pointed out already, many scientists include both the groups under a common division. Another evidence to support the bacterial nature of blue greens is the absence of true algal characters in them. Blue greens do not have a definite nucleus and plastids that are characteristic of all algae. Stainer *et al* (1971), regard, blue greens as bacteria because of similar biochemistry and cellular organization. Round (1973), treats blue greens under large algae but points out that both the groups are more primitive and have mutual affinities even though divergent in certain respects.

In.view of the facts mentioned above, bacteriologists have included the blue greens under bacteria, and have named them as cyanobacteria. But the liberation of O_2, during photosynthesis and the presence of chlorophyll in blue greens are typical algal characters not heard of in bacteria.

Classification

Manyphycologists including G.M. Smith (1938), divide the class cyanophyceae into 3 orders namely :

(1) Chroococcales : Cells are solitary or colonial, reproduction is by cell division. Includes 35 genera and 350 sps. Ex: *Gloeocapsa.*

(2) Chamaesiphonales: Members are generally epiphytic, cells are either solitary or gregarious tending towards a filamentous habit, reproduction by spores. Includes 30 genera and 150 sps. Ex: *Chamaesiphon.*

(3) Hormogonales: Members are filamentous reproducing by hormogones. Reproduction may also take place by akinetes or heterocysts or by both, includes about 100 genera and 100 sps. Ex: *Nostoc, Rivularia, Oscillatoria, Scytonema* etc.

Fritsch (1935), recognizes 5 orders in the class Myxophyceae. They are:

(1) Chroococcales (2) Chamaesiphonales

(3) Pleurocapsales (4) Chamaesiphonales

(5) Stigonematales

In this book the following members are discussed :

1. Microcystis **2. Anabena**

3. Spinilina and **4. Scytonema**

CYANOPHYCEAE (Cyanobacteria)

CHROOCOCCALES

CHROCOCCACEAE-
MICROCYSTIS

Occurrence : *Microcystis* is a fresh water planktonic alga growing in ponds, lakes, tanks, rivers and stagnant rain water pools. It is also found occasionally in salt lakes. The plant body which is unicellular either may be planktonic, or form dense mats of waterblooms. Excessive growth of this alga sometimes covers the entire surface of the pond. Some species of *Microcystis* have been reported from rice fields and moist soils rich in lime. A few forms have also been reported to occur on the wet barks of trees.

Desikachary (1959), has reported 18 species of *Microcystis* from India of which *M.flos-aquae* and *M. aeruginosa* are of frequent occurrence *(ocystispseudofUamentosa).*

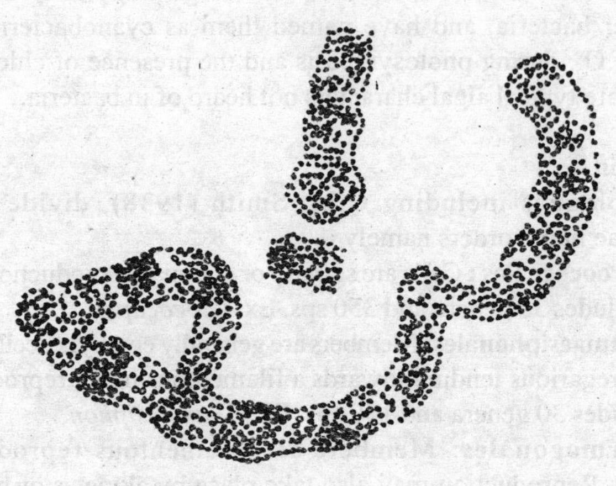

Fig. 8.2 Cyanobacteria
Microcystis pseudofilamentosa

Structure of the thallus : Unicellular members aggregate themselves into colonies. The colonies may be shapeless, spherical, ellipsoidal, elongated or clathrate *(M. aerugionsa).*

The cells are usually spherical in shape but in some species they may be elongated or ellipsoidal. A common mucilaginous matrix packs all the cells together. The mucilage sheath is hyaline and can be demonstrated only with special staining technique.

Cell structure : In its general structure the cell is typical to acyanophycean cell. Numerous psendovacuoles or gas vacuoles are present in the cell. The vacuoles are irregular in shape, and their occurrence is subject to certain ecological conditions.

Reproduction : Cell division is the only means of reproduction and has been reported in all the species. Cells divide in all the plants. In elongated cells, the plane of division is transverse, and a part of the colony separates to produce new colonies.

In *M.flos-aquae,* reproduction is reported to occur by means of nannocytes.

Water blooms : Among the Indian species, *M. aeruginosa* produces water blooms in stagnant waters particularly in the temple tanks of South India. It emits foul odour and thus spoils the water. *M. proctocystis* grows in the receded waters of rivers such as Cauvery and Ganga. Excessive growth of *Microcystis* is also responsible for the death of freshwater fish. It also grows in water reservoirs making the water non Potable.

CYANOPHYCEAE
NOSTOCALES
OSCILLATORIACEAE
SPIRULINA

Occurrence : *Spimlina* is a free floating alga found in fresh water ponds and ditches. It is one of the common planktonic alga. The alga may also grow in brackish water containing high concentration of salts. Some of the common Indian species are *S.major & subtilisima,* and *S. subsala, S. platensis* etc.

Structure of the plantbody : The plantbody *of Spimlina* is a trichome and formed by a single long spirally twisted cell. Sometimes, cross walls may be found making it multicellular. According to some phycologists, the cross walls are incomplete. The spiral cell twisting of the trichome may be loose or close, and the cells are cylindrical in shape and are without sheath. The terminal cells are rounded and is without calyptra. The cells exhibit active rotational movement.

Cell structure : Generally there is no sheath. Cell wall is multilayered and chemically it is made up of nucopolymer and pectic compounds. When sheath is present, it is made up of polysccharides. In the cell wall, cellulose is absent. There are no chloroplasts. The thylakord lamellae are distributed randomly in the cell. Chlorophylla is present. There i s no chlorophyll *b.* Apart from chlorophyll other pigments are–carotenoids, phycocyanin and phycocerythrin. True vacuoles are not present in the cell pseud–Ocacuoles or gas vacuoles present help to regulate buoyancy.

As is true to cyanophycean cells, the nucleus is prokaryotic.

Reproduction: No sexual reproduction. Asexual reproduction occurs by fragmentation.

Economic Importance

1. *Spirulina is* a N fixing blue-green alga.
2. *Spirulina* causes water blooms.

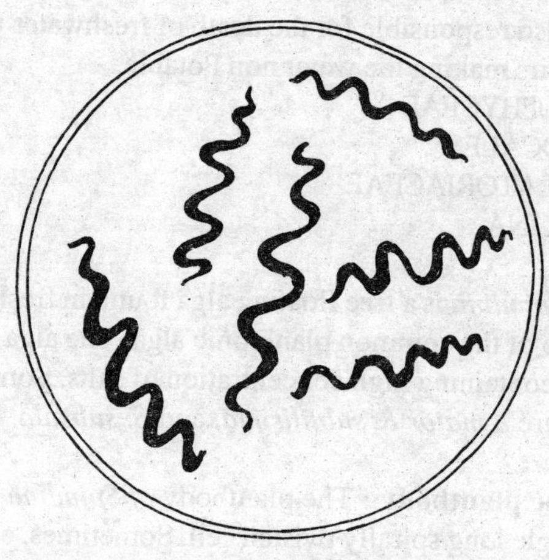

Fig. 8.3 Cyanobacteria
A diagrammatic view of Spirulina platensis

3. Some of the current researches indicate that *Spirulina* holds a great promise as a source of single cell protein (SCP). SCP is the protein obtained from unicellular organisms. Even though much research work has been conducted on the single celled alga *Chlorella* as a source of food, research

on *Spirutina is* still in its infancy Biotechnology section of Central Food Technological Research Institute has done a great deal of research on exploiting *Spirulina* as a source of food. According to CFTRI, this alga probably has the highest protein content (60-65%) as compared to other algal forms.

Spirulina can be cultured in clean water, brackish water and also in sewage systems. The large filamentous nature of the trichomes makes harvesting easy. Besides its use as protein source for food *Spirulina* has B–carotenes and vitamin B–complex particularly, cobalamine. *Spirulina* has become popular in the western countries as a source of natural or health food.

Spirulina can also be used as animal and poultry feed. Among the therapeutic uses *Spirulina* is known to possess cholesterol lowering properties. *Spirulina* has become popular in the west as a swimming food. Tablets prepared from spray dried *Spirulina* has been used as an appetite suppresant.

In our country research on *Spirulina* is being carried out in many centres other than CFTRI. These are– Navasari-in Gujarat; National Environmental Engineering Research Institute, Nagpur; National Botanical Reserach Institute, Lucknow; Durgappa Chetiar Research Centre, Madras etc. Outside India, research centres engaged in *Spirulina* production are French Petroleum Institute Mexico, Institute for Desert Research; Israel etc. In conclusion, it may be said that *Spirulina* holds a great potential not only as a source of food, but as some researches indicate, also to treat sewage and effluents which have a great role to play in pollution control.

CYANOPHYCEAE
NOSTOCALES
NOSTACACEAE
ANABENA (GR. *ANABENA =l* WILL GO UP)

Occurrence : This blue, green alga grows in moist rice fields or temporary puddles with shallow water. Many species are exclusively planktonic in lakes ponds and impart a blue green colouration to the water. Some times profuse growth of this alga leads to water blooms.
Some species of *Anabena* are symbiotic, found in association with higher plants like *Azolla-a pteridophyte (A.azollae)* and *Cycas* a gymnosperum *(A-cycadacearum)*. The alga has the capacity ofN fixation and in turn is provided shelter from the higher plants. Alga is said to *be endophytic* as it is found inside another plant.

About twenty five species are recorded form India, of which the important ones are–A *planctorica A.Spiroides, A.aphanozomenoides, A.sphaerica* etc. Desikachary (1959), considers *A.oscillatoroides* as the type species.

Structnre of the thallus : The plant body is a filament or trichome either present singly as a planktonic form-or trichomes may join together in a mucilaginous matrix of various shapes. But in some forms mucilage sheath surrounding the trichome is absent. The trichomes are more or less straight and not contorted. Colonies of *Anabena* are always smaller than those of *Nostoc.*

Trichomes of *Anabena* are made up of spherical, or barrel shaped, or subcylindrical cells. Trichomes are unbranched, uniformly broad throughout, or only the apices are attenuated (In–*Nostoc,* cells are bead shaped). Cells have a typical cyanophycean structure. A number of gas vacuoles are found more particularly in planktonic species. The cytoplasm is either homogenous or grannulated, and the cell contents have shades of blue green.

Between the vegetative cells are found two types of specialized cells. These are akinetes and Heterocysts. Akinetes are long (4 to 5 times longer than the vegetative cells), thick walled cells that help in reproduction. Heterocysts are also thick walled cells, but not as long as akinetes. They are distributed terminally or more often in between (intercalary) the vegetative cells at regular intervals. In some species akinetes occur adjacent to heterocysts, but in still others, akinetes develop in long chains away from heterocysts.

Reproduction : *Anabena* reproduces only asexually. The following are the methods of asexual reproduction :

Hormogones : The filament (trichome) frequently breaks into smaller pieces called hormogones. Usually the breakage occurs at the region of heterocysts. A hormogone may be defined as a portion of the trichome in between two heterocysts. The hormogones remain motile (gliding movement), and at a later stage develop into long filaments.

Akinetes : These are long thick walled cells that are found in between the vegetative cells either isolated, or in chains. The number and position of akinetes is a specific character. Further, the akinetes may be on one side of heterocyst or on both sides, or away from it. Akinetes are generally cylindrical, but they may be spherical also, as seen in some species. Larg (1965), has worked on the electron microscopic structure of akinete and its development in *A-azollae.* According to him the sequence of development is as follows :

(i) Formation of a bilayered envelop
(ii) Contortion of photolamellae

(iii) Loss of granular inclusions
(iv) Change in nucleoplasmic material
(v) Concentration of photolamellae towards the poles.

Nitrogen deficiency in the surroundings seems to be a factor contributing to the development of akinete. There are reports that akinetes participate in N fixation.

The akinetes get separated from the filament and germinate to form a few celled filament (hormogonium).

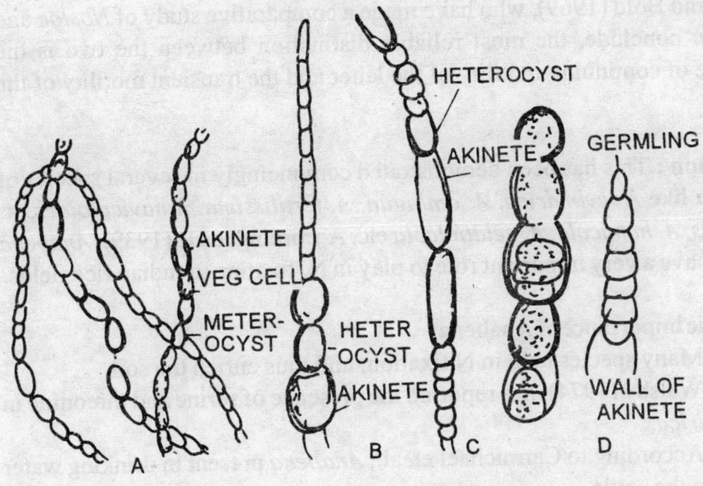

Fig. 8.4 Cyanobacteria
Anabena A.B.C. Filaments with heterocysts and akinetes. D. Germination of akinetes.

Heterocysts : These are slightly larger than vegetative cells. Heterocysts have a thick, two layered envelop with the inner wall forming a knob like projection into the cell cavity. The knob like projections are called polar nodules. There is a single polar nodule in a terminal heterocyst (towards the

filament, while there are two polar nodules in an intercalary heterocyst. While the heterocysts are mainly involved in N_2 fixation, there are occasional reports of heterocyst germinating and developing into a new filament *(A. naviculoides, A. cylindricae).* On germination of a heterocyst, a hormogonium develops first, and later it develops into a new filament.
(Geilter. 1921, Pandey. 1965). During germination the pigments reappear and the cell contents transform into a germling.
Spratt (1911), has reported the development of endospores in *A. cycadecae.*

Differences between Nostoc acid Anabena

	Anabena	Nostoc
1.	Trichomes are not contorted generally.	Trichomes are contorted.
2.	Firm mucilage sheath does not surround trichomes.	A firm mucilage sheath is present .

Krantz and Bold (1969), who have made a comparative study of *Nostoc* and *Anabena* conclude, the most reliable distinction between the two is the presence of continuing motility of the latter and the transient motility of the former.

N_2 fixation : This has been demonstrated convincingly in several species of *Anabena* like A. *cylmdrica, A. ambigua, A. fertilissima, A.naviculoides, A. variabilis, A. humicola, A. gelatinlcola* etc. According to De (1939), *Anabena* species have a very important role to play in N, fixation in indian rice fields.

Economic importance of Anabena

1. Many species help in N fixation, and thus enrich the soil.
2. Walsby (1974), has reported the presence of serine and threonine in *A. cylindrica.*
3. According to Carmichael et. al., *Anabena* present in drinking water is toxic to the cattle.

CYANOPHYCEAE
HORMOGONALES(NOSTOCALES)
SCYTONEMATACEAE
SCYTONEMA

Occurrence : Generally found as intricately fused filamentous masses, either in purely aquatic or semi terrestrial conditions. Species of *Scytonema* like *S. simplex,* are. completely submerged forming the undergrowths of ponds and lakes. Some species are found attached to aquatic plants. Terrestrial species like *S. oscillatum,* form mats on wet rocks, damps, barks of trees etc. Some species of *Scytonema* are also known to occur as "Phycobionts" (algal partners), in Lichens.

Structure : The plant body is a filament and is surrounded by mucilage sheath, like in other blue-green algae. The sheath is lamellated and of golden yellow or brown color. The filaments possess either intercalary or terminal heterocysts.

The genus is characterised by its unique type of branching. This is known as "false branching" or "Gemminate branching".

The filament breaks up near a heterocyst and the broken end protrudes out of the sheath as a branch. If both the free ends protrude out as branches there will be two branches and it is known as "gemminate branching". If only one develops into a branch, it is known as "false branching". Later the branches develop their own mucilage sheath.

Cause of Branching : Branching is the result of the following causes:

(1) By the degeneration of Intercalary cell : When one intercalary cell dies, the free ends of cells continue growth resulting in gemminate branching.

(2) By the development of separation discs : An intercalary cell in the filament becomes dark in colour and thin walled losing its contents. Gradually it becomes bi-concave in shape and degenerates ultimately. Soon after, branching follows, producing a pair of branches.

(3) By the formation of Loop : Geitler, for the first time in 1936, described the formation of loops resulting in branching of *Scytonema.* The filament bulges out and forms a loop. One of the terminal cells in the loop degenerates and this results in two pseudobranches.

(4) By the Heterocysts: "The trichome breaks up at the point of heterocyst" and branching follows the usual pattern.

Cell structure : Cells are cylindrical or rectangular in shape. Heterocysts are found in between the vegetative cells. The terminal cell is usually hemispherical.

There is no sexual reproduction

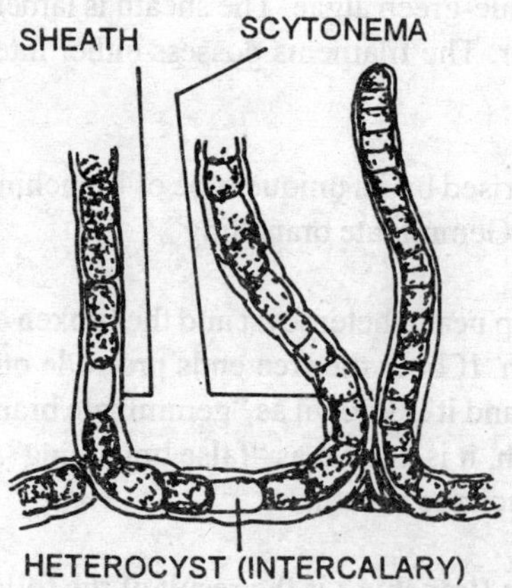

Fig. 8.5 Cyanobacteria
Scytonema. Habit

Asexual reproduction is brought about by the formation of hormogonia. A hormogonium is formed by the formation of two separation discs. Each one consists of several cells and germinates into a new filament.

A few species may produce Akinetes also (Burzi, 1879).

Economic importance of Blue-green Algae
Blue-green algae are the most ancient algae, perhaps they were the first green organisms to appear on earth, as such, they form a very important link in the photosynthetic food cycle. They form food for fish and other aquatic animals which are in turn consumed by other organisms at the higher trophic level.

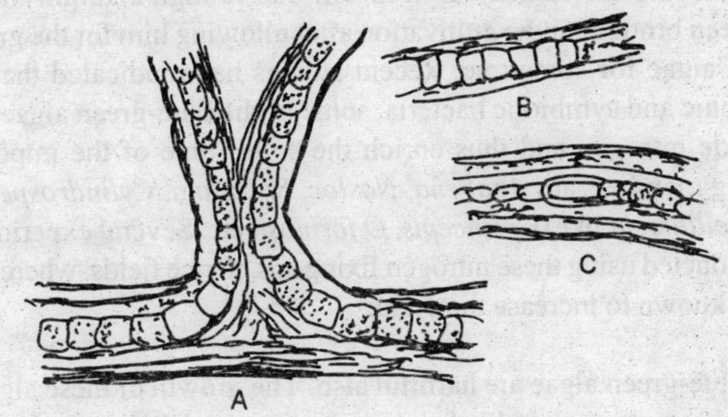

Fig. 8.6 Cyanobacteria

A. A portion of the plant showing false branching **B.** A few cells convered with mucilage

C. A part of the branch showing heterocyst

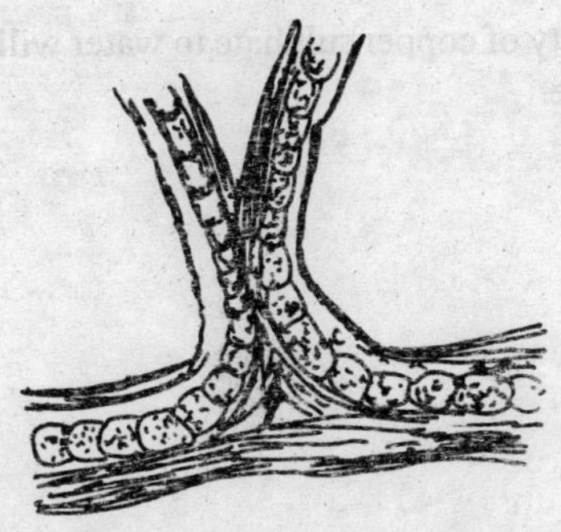

Fig. 8.7 Cyanobacteria

Origin of false branching

Some species of blue-green algae including *Nostocare* are boiled in soups and utilized as food. Some of the semi-terrestrial forms add organic matter to the soil and increase the fertility. Barren soil due to high alkalinity in our country has been brought under cultivation after allowing him for the growth of blue-green algae for sometime. Recent studies have indicated that like some autolrophic and symbiotic bacteria, some of the blue-green algae, also fix atmospheric nitrogen and thus enrich the soil. Some of the important nitrogen fixing members are *Anabena, Nosloc, Spirulina, Cylindrospemum* and sps. of *Oscillatoria* like 0. *princepis, O.formosa* etc. Several experiments have been conducted using these nitrogen fixing sps. in rice fields, whereupon the fertility is known to increase the yield.

Some of the blue-green algae are harmful also. The growth of these algae in abundance makes water unsuitable for consumption as such they are harmful when they grow in water reservoirs. Many a time they produce foul odour (pigeon odour), also making the water distasteful. Excessive growth of some of these algae like *Aphanizomenon* may result in the death of fish due to depleted oxygen' supply. Species of *Microcystis* and *Anabena* may even kill the cattle which drink water polluted by these. These algae are known to cause various liver, heart and lung diseases in cattle ultimately leading to death. *Microcystis* is known to produce a potent neurotoxin which acts as the active principle. If filaments of *Oscillatoria* are consumed with water they cause symptoms of conjunctivitis, itching, blocking of nasal passages and even bronchial asthama.

Adding a little quantity of copper sulphate to water will effectively check the growth of these algae.

9

MICROALGAE

GENERAL ACCOUNT

The study of Algae is called Phycology. Algae have been known to human beings since time immemorial, though their scientific study is recent. Early Chinese, Roman and Greek literatures contain numerous references to this group of plants. Ancient Hawaiians used algae for food. At present, the study of algae is one of the important fields of botanical science having far reaching implications from the point of view of human welfare.

Introduction

According to the present trend of plant classification, however, algae do not constitute any compact group. The term algae refers to about eight divisions of the plant kingdom. From the point of view of convenience, however the term has still been retained. All algae are chlorophyllous thallophytes, and hence autotrophic. They are non vascular plants with predominant aquatic distribution. There are about 30,000 species of algae.

Occurrence and Habit

The members are aquatic being disributed both in fresh and salt waters. Fresh water algae abound in stagnant waters, tanks, ponds, pools, ditches, slow running streams, rivers etc. Some of the terrestrial forms grow ubiquitously whenever nature offers a damp surface. They grow on wet trunks and branches of trees or even on rocks. They may also live as epiphytes or epizoophytes (algae growing on animals, specially molluscan shells). Freshwater algae may be found free, floating or attached to rocks, stones or other higher water plants like *Elodea, Ottftlia* etc.

Some of the algae .are endophytic growing in the company of higher plants. Some of the cyanophycean algae like *Nostoc* and *Anabena* which are known to possess the capacity of nitrogen fixation are found forming symbiotic association with plants like *Anthoceros, Isoetes, Cycas* etc.

Marine algae are commonly called sea weeds and they occupy a large area(some times several thousand miles). They may be free floating or attached. The free floating marine algae are known as "Phytoplanktons". Those which are attached, are found growing in shallow waters forming "Benthons".

Range of Vegetative Structure

The range of thallus organisation in algae may be classified into the following six types –

(1) Motile types –

This is the simplest type. Plant body is motile and unicellular throughout. In some case? motile plants may be colonial with little or no division of labour E.g., *Chlamydomonas, Volvoxetc.*

(2) Palmeooid and Dendroid types –

Palmelloid or non motile stage is found as a temporary state even in motile forms. But in forms like *Tetraspora,* this is a permanent feature. Here the cells are surrounded by a mucilage covering. Dendroid type is a slight variation of the Palmelloid type. E.g *Tetraspora, Prasinoclcuiusete.*

(3) Coccoid forms –

in these forms the locomotory organs (flagella) are lost and the cells become rounded. E.g., *Chtorococcum.*

(4) Filamentous types –

The filamentous habit is believed to have originated from motile forms where loss of motility and restriction of cell division to one plane has resulted in the filamentous types. The filaments may be free floating or attached. All the cells in ' the filament may be alike, or one of the lower cells may be modified to help in the attachment (holdfast, hapteron etc). The filaments may be branched or unbranched. E.g., *Ulothrix,Oedogonium,Cladophora* etc.

(5) Siphonous habit –

The plant body consists of tube like filaments (branched also), but no cross walls are found. Hence the entire plant body is comparable to a single cell. Cross walls appear only to separate off the dead portions of the thallus or to demarcate the reproductive structures. Ex. *Vaucheria, Caulerpa* etc.

(6) Advanced types –

They may be monosiphonous or polysiphonous filamentous types or even true. parenchymatous.

Reproduction

Algae reproduce by two methods viz –

(1) Asexual includes vegetative propagation also)

(2) Sexual

Asexual Reproduction

This is brought about by a variety of methods –

(1) *Cell Division* – Cell division as a means of multiplication of the individuals is seen in several unicellular members. E.g., *Pleurococcus* etc.

(2) *Fragmentation* – Accidental break up of the plant body results in fragments. Each fragment develops into a new individual Ex. *Spirogyra, Ulothrix* etc.

(3) *Hormogones and Hormocysts* – This type of vegetative multiplication is found in some cyanophyceae. Filaments break up into 3–4 celled bits called hormogones (thick walled ones are called Hormocysts), which develop into a new individual.

(4) *Protonema* – These are thread like bodies helping in vegetative multiplication. E.g., *Chara.*

(5) *Amylum Stars* – These are starch filled stellate structures helping in asexual reproduction.

(6) *Akinetes* – These are non motile oval shaped structures formed in cells usually to withstand the adverse environmental conditions. They are formed in many members of chlorophyceae.

(7) *Aplanospores* – These are non motile spores, formed in cells. At favourable season these develop into new individuals.

(8) *Zoospores* – By far this is the commonest method of asexual increase in algae. These are motile spores adapted to aquatic habit. They are formed.in cells *(źffosporangia),* either singly or in large numbers. They are biflagellate mostly, may be quadriflagellate or multiflagellate also. Flagella may be of equal length *(Isokontean),* or of unequal length *(stephanokontean).* Zoospores are mostly pyriform in shape. They swim for some time in water and then develop directly into a new individual.

Besides the above, various other algae reproduce asexually by means of Endospores, Autospores, Auxospores, Carpospores, Monosoores, Hyponspores, Statosporesetc.

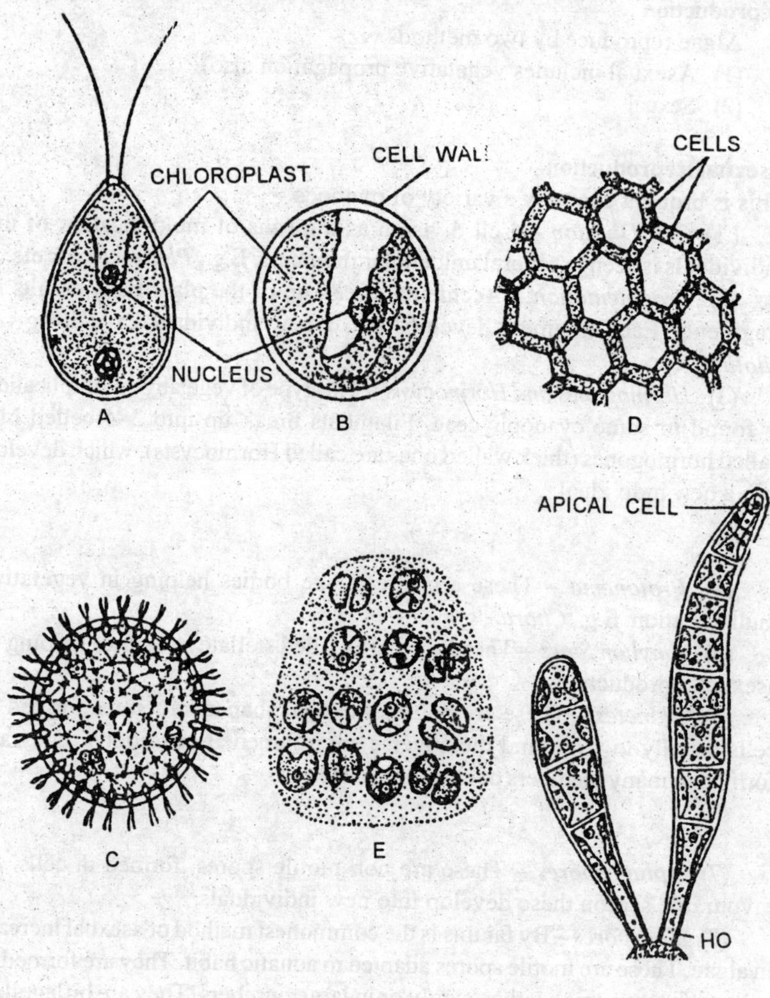

Fig. 9.1 Microalgae

Range of vegetative structure in algae

A. Chlamydomonas (unicellular, motile)
B. Chlorella (unicellular, non motile)
C. Volvox (coenobium)
D. Hydrodictyon (multi-cellular, colonial non motile)
E. Tetraspora (Palmelloid)
F. Ulothrix (unbranched filament)

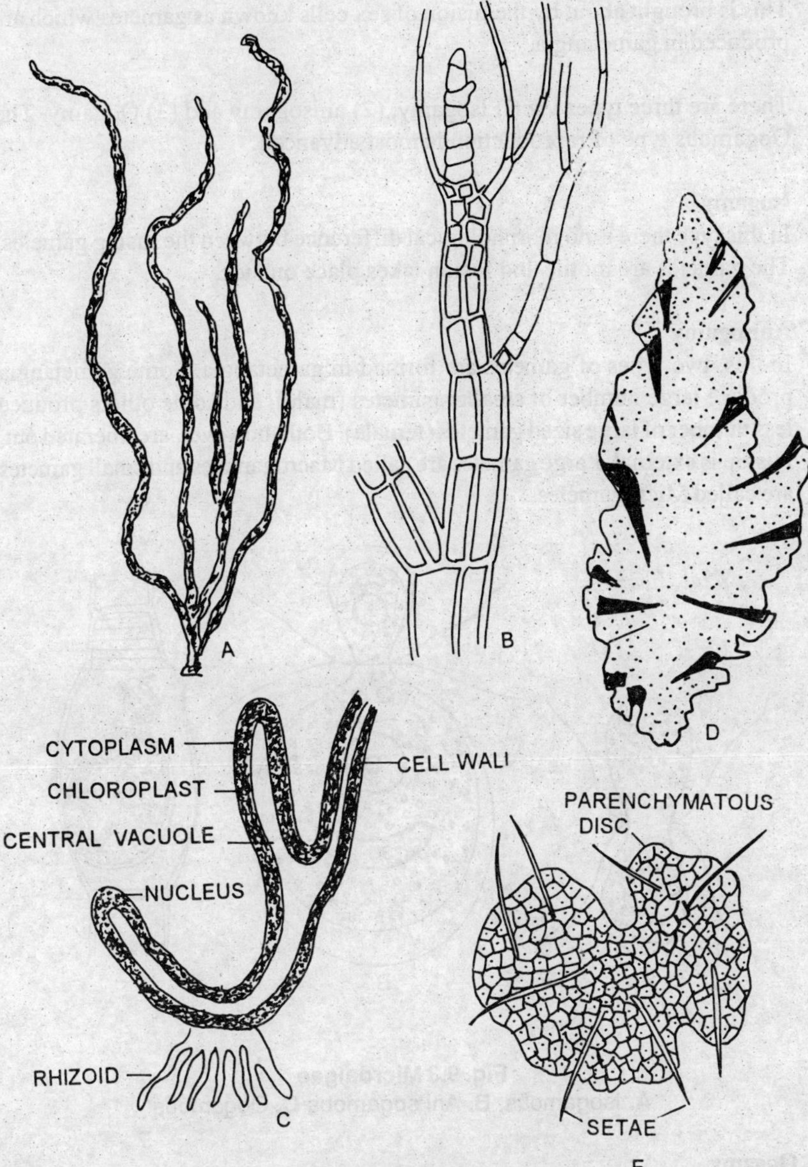

Fig. 9.2 Microalgae

Range of vegetative structure in algae

- **A.** Nemalion (multiaxial)
- **B.** Polysiphonia (Branched filament)
- **C.** Vacheria (coenocytic, spihonous habit)
- **D.** Ulva (advanced parenchymatous)
- **E.** Coleohoatae (Heterotrichous)

Sexual Reprodcution

This is brought about by the fusion of sex cells known as gametes which are produced in gametangia.

There are three types viz (l) Isogamy, (2) anisogamy and (3) Oogamy. The Oogamous type of reproduction is most advanced.

Isogamy

In this type there is no morphological difference between the fusing gametes. The gametes are motile and fusion takes place outside.

Anisogamy

In this, two types of gametes are formed in gametangia. Some gametangia produce large number of smaller gametes (male), and some others produce less number of large sized gametes (female). Both, however, are liberated out. Fusion is external. Large gametes are called Macrogametes and small gametes are called Microgametes.

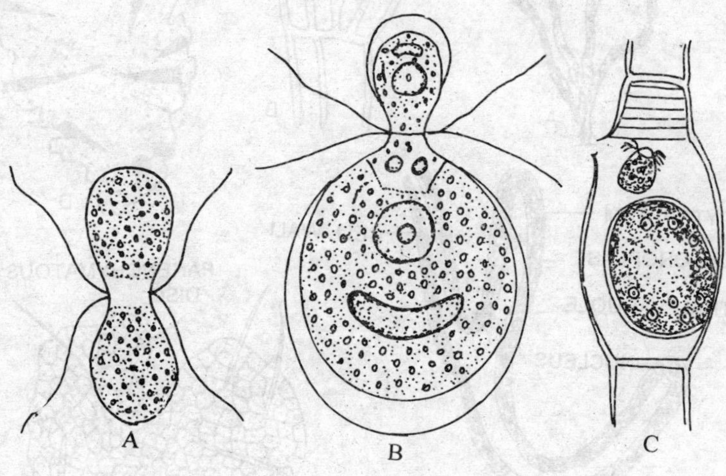

Fig. 9.3 Microalgae
A. Isogamous, **B.** Ani sogamous **C.** Oogamous

Oogamy

In this, the male and female gametangia are different. The former is called Antheridium (Spermatangium), and the latter Oogonium. Antheridium produces many flagellate gametes. Oogonium produces only one non-motile female gamete (egg), which never comes out. Fertilisation is internal.

Alternation of generations

In some algae, the life history comprises of two distinct generations viz, Sporophyte and Gametophyte. The former is asexual and diploid, the latter sexual and haploid. Haploid gametophyte reproduces sexually and the fused product (Zygote), directly develops into the diploid sporophyte (No alternation of generation if the zygote undergoes reduction division). This diploid sporophyte reproduces asexually by spores. During spore formation reduction division takes place, hence the spores are haploid. They develop into gametophyte. If both gametophyte and sporophyte are alike morphologically, alternation of generation is said to be Isomorphic, and if they are different, Heteromorphic.

Only a few algae, however, exhibit alternation of generations.

Classification (All systems have not been discussed).

Before 20th Century, only four classes were recognised in algae.

They are –

(1) Chlorophyceae (2) Phaeophyceae
(3) Rhodophyceae and (4) Cyanophyceae

Fritsch divided Algae into eleven classes namely

(1) Chlorophyceae (2) Xanthophyceae
(3) Chrysophyceae (4) Bacillariophyceae
(5) Cryptophyceae (6) Dinophyceae
(7) Chloromonodineae (8) Euglenophyceae
(9) Phaeophyceae (10) Rhodophyceae
(11) Myxophyceae

In the classification proposed by Fritsch, importances given to pigmentation, nature of plant body, and reproductive features, metabolic products, locomotory organs etc.

Gilbert M. Smith (1938), classifies algae as follows –

Division	Class
(1) Chrlorophyta	(1) Chlorophyceae
	(2) Charophyceae
(2) Euglenophyta	(1) Euglenophyceae
(3) Pyrrophyta	(1) Desmophyceae

	(2) Dinophyceae
(4) Chrysophyta	(1) Chrysophyceae
	(2) Xanthophyceae
	(3) Bacillariophyceae
(5) Phaeophyta	(1) Isogenerate
	(2) Heterogenerate
	(3) Cyclosporeae
(6) Cyanophyta	(1) Myxophyceae
(7) Rhodophyta	(1) Rhodophyceae

Algae of uncertain systematic position –

1. Chloromonadales
2. Cryptophyceae

Papenfuss divides algae into eight phyla. They are-

1. Chlorophycophyta
2. Charophycophyta
3. Euglenophycophyta
4. Chrysophycophyta
5. Pyrrophycophyta
6. Phaeophycophyta
7. Schizophyta
8. Rhodophycophyta

Champman (1926), proposed a system of classification of algae based on morphological characters, pigments and biochemical difference. He divides algae into four –

1. Euphycophyta
2. Myxophycophyta
3. Chrysophycophyta
4. Pyrrophycophyta

He includes Charophyceae, Chlorophyceae, Phaeophyceae and Rhodophyceae under Euphycophyta.

In this book, only Chlorella and Diatoms are discussed.

CHLOROPHYTA

Order : Chlorococcales

The order chlorococcales includes unicellular coenocytic or colonial non motile green algae of the class chlorophyceae. Colonial cells have definite number of cells and the colonies are referred to as coenobia.

The order has been divided into a number of families such as chlorellaceae, Hydrodictyaceae etc. *Chlorella* belonging to chlorellaceae is discussed in this book.

CHLOROPHYCEAE

CHLOROCOCCALES
CHLORELLACEAE
CHLORELLA

Chlorella is a unicellular green alga found in fresh water ponds and ditches, in moist soil, damp surfaces such as surface of tree trunks, pots, damp walls etc. According to Kessler et al. (1968), *Chlorella* occurs in marine waters also. The genus is divided into eight species using morphological and physiological criteria (Chapman, 1973). Some of these like *C. vulgaris, C. gongtomerats, C. conductix* and *C. parasitica* are reported from India. Species of *Chlorella* may be free living, symbiotic or even parasitic. *C. vulgaris* and *C. variegata* are free living. Symbiotic species *of Chlorella* may be endophytic or endozoic. Symbiotic endophytic species is *C. lichina,* found in the lichen *Celicmm chlorina.* Endozoic chlorella is called *zoochlorella (C. conductix),* is found in *Hydrastentor, Paramecium* etc. Maheswari and Gupta (1962) have reported *C. ellipsoidea* occuring on wet cotton fabric in Kanpur. *C. parasitica* is a parasite found in *Ophyridium* and sponges. There are some reports of saprophytic *Chlorella* also under culture conditions.

Structure of the plant body

Chlorella has a unicellar non-motile plant body. The cells are small (range in size from 2–12 mm), globular or ellipsoidal in shape. The cell wall is three layered–outer mucilaginous, middle pectic and inner cellulosic. Occupying most part of the cell is, a parietal cup shaped chloroplast with a pyrenoid. In some species pyrenoid may be absent. Lying in the cavity of the chloroplast is the cytoplasm, with a single nucleus suspended in it. Flagells, stigma and contractile vacuoles are absent.

Ultra Structure

Electron microscope studies have revealed that internal to the cell wall is the two unit cell membrane enclosing cytoplasm and the chloroplast. Much of the cell interior is occupied by the chloroplast which is 'C' shaped. There are a number of thylakoid lamellae, but these do not have the stack like (grana) organisation except perhaps in *Chlorella pyrenoidosa.* The chloroplast is also enclosed in a double membrane which is distinct from hyaloplasm. The pyrenoid when present is bordered by numerous starch plates. The cell has a single nucleus suspended in the cytoplasm. The nuclear envelope is double layered and porous. Other cell inclusions of the cell are mitochondria, golgi bodies and a few vacuoles. Endoplasmic reticulum and cytoplasmic lipid granules also have been reported in some species of *Chlorella* (Silverberge and Sawa, 1974). Atkinson *et al.,* (1972), have reported the presence of sporopollenin in the cell wall of *Chlorella.*

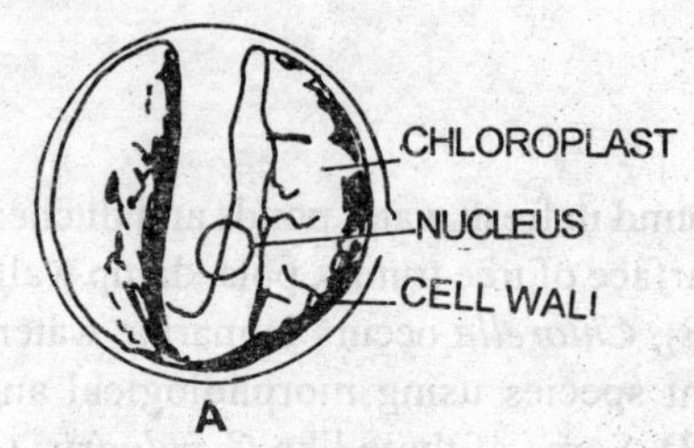

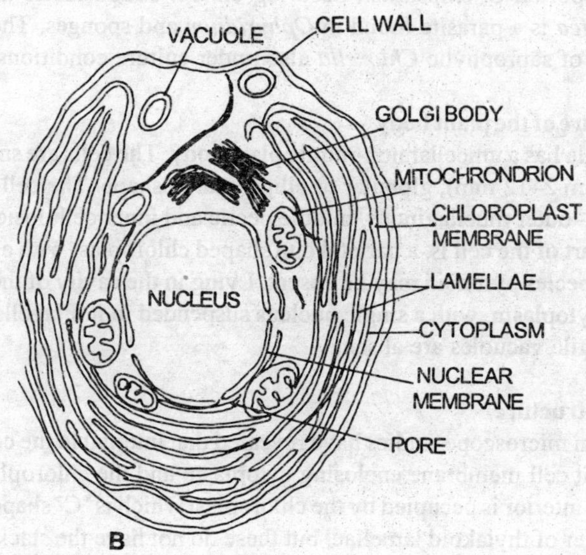

Fig. 9.4 Microalgae

A,B. Chlorella, A,A. Cell under light microscope

B, A. Cell under electron microscope (diagrammatic)

Reproduction : No sexual reproduction has been reported so far in *Chlorella*. The only mode of reproduction is asexual. This is brought about by the formation of *autospores*. Spores are' produced as a result of cell division. Each cell divides to form two, four or sometimes sixteen daughter cells.These daughter cells are non-motile and have the same shape as the parent cell but are smaller in size. The parent cell wall ruptures releasing the autospores. Eventually the autospores mature into adult *Chlorella* cells. Bisalputra and Weier (1966), Wanka (1968), and Griffiths and Griffiths (1969), have studied the auto spore formation under electron microscope. Presence of sulphur and sometimes nitrogen is a prerequisite for the formation of autospores.

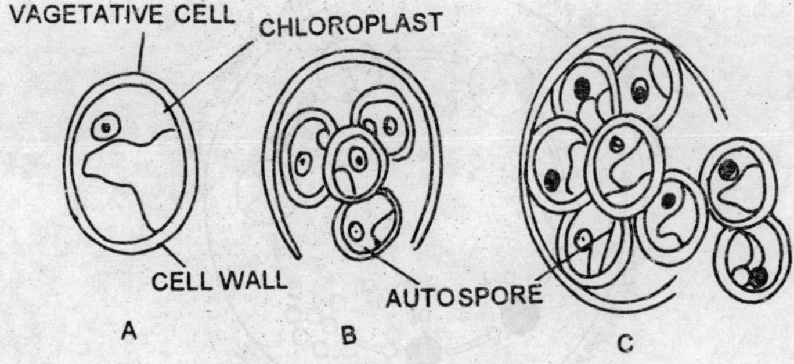

Fig. 9.5 Microalgae
Chlorella, **A.** Vegetative cell, **B,C.** Autospore formation

The most characteristic feature in asexual reproduction is that even though

Chlorella is aquatic, *no swarmers are produced during reproduction.* The report of Bendix (1964) that *Chlorella* produces motile cells which could be gametes needs confirmation.

Physiology of Chlorella

Cells of Chlorella can he easily cultured on a number of media including salt solution (0.2 m). Glucose Agar medium is the most suitable according to Atari et al. Some species of *Chlorella* seem *to* exhibit saprobic tendencies and grow luxuriantly on organic media even in dark. Cells cultured in dark differ from those grown under light. Quiscent dark (cells grown under darkness) cells are small and have a photosynthetic quotient (O/co,> of I). They give rise to dividing dark cells, which grown under light develop into light cells'. 'Light' cells are larger, have poorly developed chloroplast and consequently less active photosynthetically. (Photosynthetic quoteint of 3) During development the amount of DNA increases.

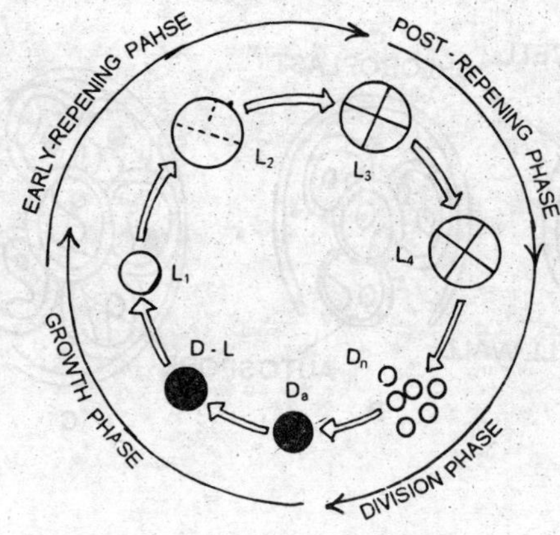

Fig. 9.6 Microalgae

Diagrammatic representation of life-cycle of Chlorella ellipsoidea (after Tamiya)

Chlorella has been grown under synchronous culture also (In synchronous cultures all the cells will be in the same state of development). Tamiya (1963), has identified four phases in the life cycle of *Chlorella* during his studies of synchronous culture of *C. ellipsoidea.* "These four phases are:

1. *Growth phase :* In this, autospores increase their size using the photosynthetic products.
2. *Early ripening phase:* Cells prepare themselves for dividing.
3. *Post-ripening phase:* Cells divide twice either in light or darkness.
4. *Division phase:* Parental cell ruptures releasing the autospores.

The photosynehetic system of Chlorella is biochemically similar to that of Angiospenns and as such it is a material for the study of photosynthesis. Otto Warburg (1919), was the first person who employed *Chlorella* for photosynthetic studies. The discoverer of CO, reduction cycle (Melvin Calvin), also used *Chlorella (C. pyrenoidosa),* for his studies. The following are the advantages of *Chlorella* in experimental studies on photosynthesis.

(1) The cells can be easily cultured.
(2) Chlorophyll pigment and the end products of photosynthesis are similar to those of Angiosperms.
(3) Protein contents are high in young cultures, but in old cultures carbohydrates are more.

Economic Importance

(1) Extensively used in experimental studies on photosynthesis.

(2) *Chlorella* is the centre of some recent researches concerning alternate source of food for human consumption. Nutritional content of *Chlorella* is very high and is cultivated in mass cultures in many countries like USA, Japan, Germany etc. The reasons for *Chlorella* being an alternate source of food are as follows :

(a) It grows rapidly in mineral solutions.

(b) Ash analysis of *Chlorella* cells show that its inorganic contents are almost on par with corn.

(c) *Chlorella* has 50% proteins, carbohydrates 20%, amino acids and many minerals.

(d) Dried *Chlorella* cells have the following amino acid content.

Arginine	2.06%
Histidine	0.62%
Isoleucine	1.75%
Leucine	3.79%
Lysine	2.06%
Methionine	0.36%
Phenylalanine	1.81%
Threonine	2.12%

Tryptoplhan	0.80%
Valine	2.47%

(e) *Chlorella* cells are also rich in vitamins A, B and C.

3. Feeding experiments with dry powder of *Chlorella* have shown that in both animals and human beings there is a general improvement of health. But *Chlorella* has not become popular due to its unpleasant odour. Pilot productions of dry *Chlorella* powder have given a yield of 20 gms per square meter, per day. This amounts to approximately 17.5 tons per acre, per year.

4. *Chlorella* may be used as a raw material in several industries that require chlorophylls for deodorant proposes.

5. *Chlorella* is used for purifying air in nuclear submarines and space ships. Astronauts feed on *Chlorella* soup. It has been calculated that one kilogram of fresh *Chlorella* can supply 25 litres of oxygen required by a 70 kg roan, and also utilize Co, released by him.

6. *Chlorella* yields an antibioic ***chlorellin*** useful against several bacteria.

7. The alga lias also been in great demand in the purification and treatment of sewage. After the primary sedimantation of solids, sewage water is let into large shallow tanks containing *Chlorella.* Photosynthetic activity of *Chlorella* releases large quantities of oxygen which is used by aerobic bacteria to digest the waste organic matter. A number of sewage treatment plants using *Chlorella* exist in USA, New Zealand etc. This would perhaps be the cheapest method of sewage treatment.

Evolutionary trends : The total absence of motile phase in the life cycle in spite of the aquatic habitat is the most significant aspect in the life history of *Chlorella.* It illustrates the first step in the tetrasporine line of evolution of the thallus in algae.

CHRYSOPHYTA
BACILLARIOPHYCEAE
(Diatoms)

The Bacillariophyceae are commonly called *'Diatoms'*. They include a large number of unicellular and colonial algae, which vary greatly in form and size. They are most widely distributed among all the algae. The primary distinguishing feature is that the cell wall is silicified and consists of 2 halves, which fit together as do the two parts of petridish. Within a cell, there are one to many chromatophores.

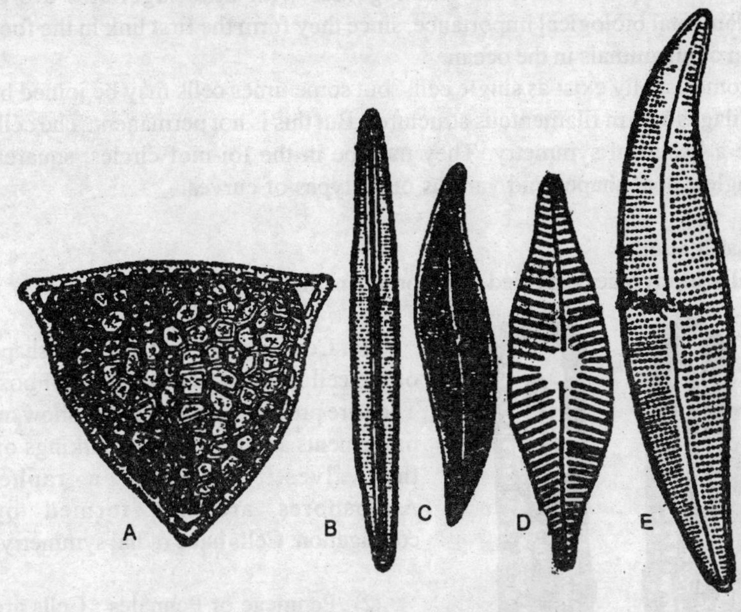

Fig. 9.7 Microalgae
A. Triceratium, B. Amphipleura, C. Pleurosigma, D. Anavicula, E. Cymbella

Reproduction is by cell division and auxospore formation. Auxospores are usually zygotic in nature produced by the union of gametes but may not be always so.

There are about 170 genera and 5,500 species majority of which are living but some known only in fossil conditions.

Occurrence

Certain genera are found only in fresh water and certain others only in salt water. Many are found in both the habitats. Many fresh water forms abound in large numbers during spring or autumn months. They seem to flourish very well wherever there is a damp surface. They are often seen as yellow oily scum on the soil surface. The marine forms of diatoms constitute a predominant part in the planktonic algae. They are so small that as many as 35,5000,000 diatoms may be present in one cubic meter of water. Many grow as epiphytes attached to the surface of higher plants and other algae.

The marine planktonic diatoms together with dinoflagellates are of fundamental biological importance, since they form the first link in the food chain of all animals in the ocean.
Diatoms usually exist as single cells, but some times cells may be joined by mucilage to form filamentous structures. But this is not permanent. The cells have a beautiful symmetry. They may be in the foi-mof circles, squares, triangles, boat shapes and various other types of curves.

Classification
Bacillariophyceae is divided into 2 orders namely,

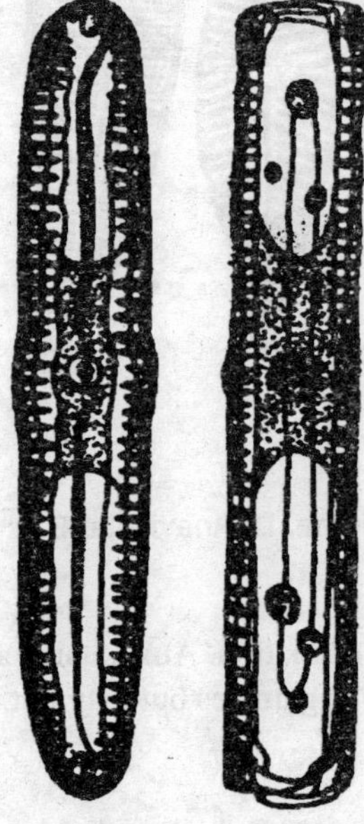

Fig. 9.8 Microalgae
A. Valve view or top view,
B. Girdle view

(1) Centricae or Centrales : The shape of the cell is equal to a circular pill box. They are predominantly marine, show no movements and bear radial markings on the valves. They lack a raphe. Auxospores are not formed by conjugation. Cells have radial symmetry.

(2) Pennicae or Pennales : Cells are elongated and boat shaped in outline. They are predominantly fresh water forms show locomotion due to streaming movement of protoplasm. They are bilaterally symmetrical.

Cell structure
Structurally the diatom cell consists of 2 parts or halves which fit in like the lid and the box. The bigger half is known as the *Epitheca* and the smaller half is known as the *Hypotheca.* The cell size is about 140μ X 20μ in Pennales (*Pinnularia).*

Each valve or half consists of a flat top (Valve face) and the margin or *girdle,* which represents the overlapping face. A diatom cell looks different when viewed from different angles. The view from the side is called the *girdle view* and the view from the top is called the *valve view.*

In the valve view, only one valve can be seen. The wall can be clearly seen and is called the 'frustule'. The cell wall is silicified and made up of pectin.On the wall, can be seen prominent markings or *costae,* extending from the margin but not reaching the centre. These costae in the wall are composed of rows of dots representing cavities. The row of costae are called *striae.* These are radially arranged in centrales. Along the centre of the valve is present a long slit in the wall which looks like a line. This is called *Raphe* and a sigmoid in shape. At the centre of the raphe is a swelling *cailedcentraliwdule.* At either end of the raphe are also present two *polar nodules.* The mucilage exudes out of the wall through these openings and supposed to help in locomotion.

Inside the silicious wall is the cytoplasmic lining, the primordial utricle surrounding a large central vacuole. Distributed in the cytoplasm, are the chromatophores–many in centrales, and one or two in pennales. They have the principle pigmentxanthophyll and carotene. The nucleus lies in the bridge like mass of cytoplasm which extends across the vacuole in the centre.

Locomotion

The centric forms are non motile. Many pennales exhibit jerking movements. These take place only in forward and backward direction along the longitudinal axis. While there is no unanimity regarding the mechanism of locomotion in diatoms among phycologists, Muller's theory has been accepted by many. According to him, movement is brought about by streaming movement of cytoplasm established along the region of raphe and nodules.

Reproduction

Generally occurs during the night. The commonest method is cell division and occasionally by auxospores formation.

Cell division

During division, the frustule slightly increase in size. This is followed by the mitotic division of the nucleus and the gradual fission of the protoplast along the value face. The two valves separate and each valve develops a new valve over the naked portion of the protoplast. The newly formed valve will always be hypotheca irrespective of the fact whether the daughter cell receives the hypotheca or epitheca from the parent cell (fig. 13.10).

It will be thus evident that if cell division continues for several generations, along one line, the size of the cell goes on decreasing. Theoretically, this progressive diminition of size is known as "Mcdonald Ptitzer law". This decrease of size, however, cannot continue indefinitely.

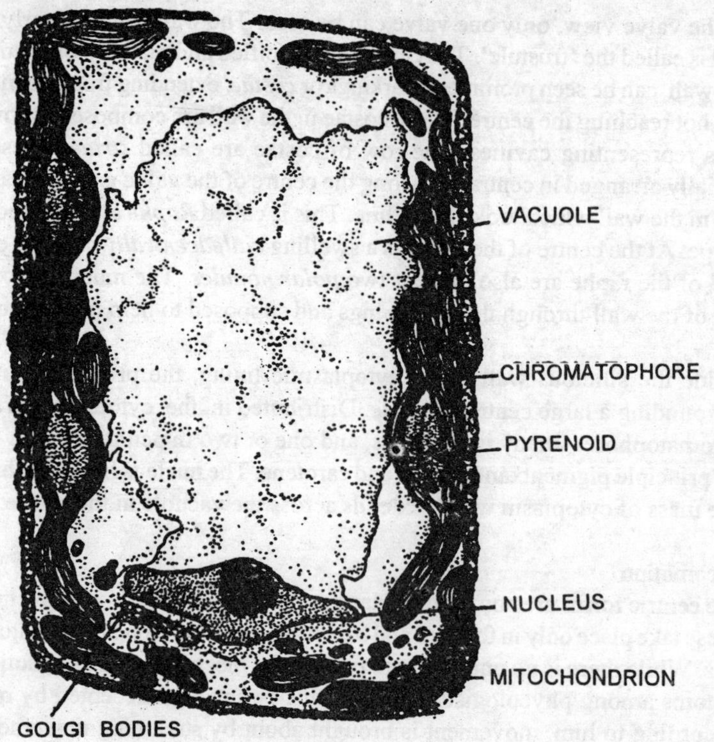

Fig. 9.9 Microalgae
Diagrammatic Sketch. Electron micrograph of a diatom cell (Melosira)

Auxospores
This is a remedial measure for the decrease in size due to continued cell division. Auxospores of most pennales are zygotic in nature and result from gametic union. But in some cases they are also formed parthenogenetically. For a long time, the auxospores of centrales were thought to be formed asexually. But it its quite probable that they are also zygotic in nature.

Auxospore formation involves a liberation of protoplast from the enclosing wall, a significant enlargement of the liberated protoplast and secretion of a bipartitesilicified wall around the protoplast. This is the auxospore. This divides to form 2 vegetative cells whose size is very near to the maximum size of the species.

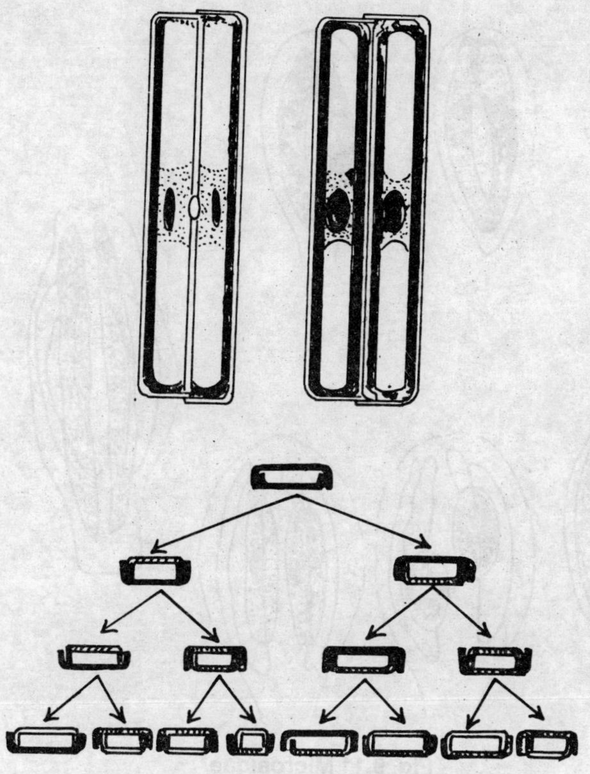

Fig. 9.10 Microalgae
Pinnularia. Stages in the cell division

Auxospores of Pennales

Auxospores are formed by any one of the following 5 ways:

(1) 2 cells conjugating to form a single auxospore.

(2) 2 cells conjugating to form 2 auxospores.

(3) 2 cells becoming enveloped into a common envelop but each giving rise .to an auxospore without conjugation.

(4) One cell giving rise to one auxospore.

(5) One cell giving rise to 2 auxospore.

The first two of the above are obviously sexual. The last three methods appear to be a sexual. But cytological studies have shown that they could also be sexual.

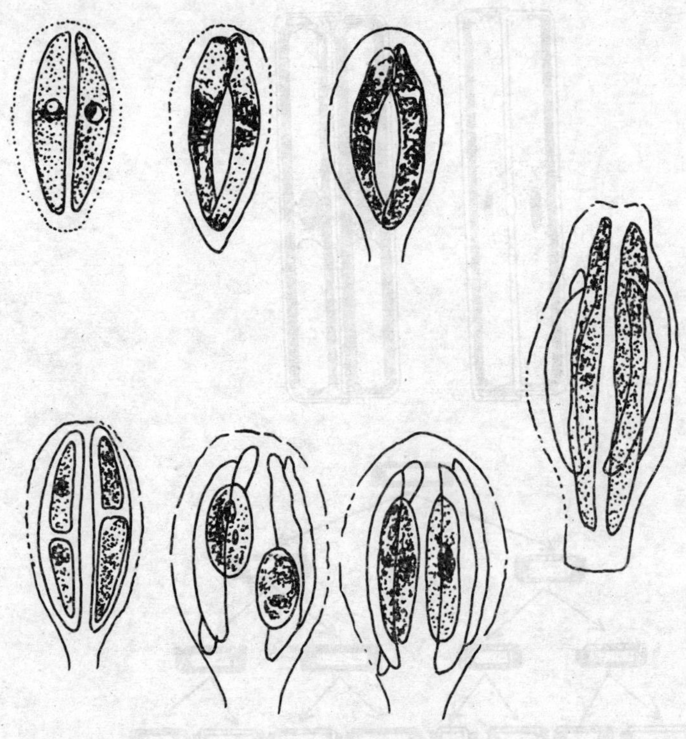

Fig. 9.11 Microalgae
Cymbella lanceolata Formation of auxospore

Auxospores of Centrales

Auxospores have been reported from a wide variety of centrales, and in all the cases, there is the formation of a single auxospore by an old or recently divided cell. As has been said already, auxospores of centrales were thought to be sexual in nature but nuclear division in connection with auxospore formation questions their interpretation. Auxospore formation is of the following types:

(1) Autogamy : In this type, the two valves of a diatom cell break apart, the diploid nuclues divides mitotically to form four nuclei of which two degenerate. The other two fuse to form a diploid nucleus. This is called Autogamy. The protoplast escapes from the parent frustule functioning as an auxospore. This later develops into a diatom cell of normal size.

(2) Oogamy : Auxospore formation by sexual process has been found in *Melasira variaus, Cycloella tennistriata* etc. Here the diploid diatom cells undergo reduction division arid form after a number ofmitotic divisions small uninucleate protoplasts ranging in number from 4 to 128. They may be uni or biflagellate. These were formerly regarded as 'Microspores'. But now they have been known to be gametes.

The non-motile gamete is produced singly by the female cell. Here, the diploid nucleus undergoes redaction division forming 4 nuclei, of which 3 degenerate. The surviving nucleus forms the egg. After fusion with the sperm, the zygote is called the auxospore. This later forms a new diatom cell directly.

Some centrales also reproduce by thick walled resting spores called 'statospores'. They are most frequently encountered in marine planteons.

Economic importance of Diatoms

Diatom cells form an important link in the food cycle. It has been estimated that of the world's total phetosynthesis, about 90% takes place in oceans by marine planktons. In this, diatoms are most prominent. They form food for marine forms, which in turn are consumed by human beings.

Fossil Diatoms

Even after the death of diatom cells, the silicious cell wall remains unaltered; millions of these accumulate at the bottom of the ocean and become hardened. This fossil deposit is known as the 'Diatomaceous earth' and is of great importance economically.

Some diatomaceous earths originated in fresh water and some in marine water. Some marine deposits are found inland due to geological changes that has changed the position of land masses and water several times. The thickest deposits discovered arein Santa Maria oil fields, California, USA measuring more than 3,000 feet The largest deposits in extent are in Lompoc, California where the beds extend over miles and over 700 feet deep.

Uses of diatomaceous earth

Diatomaceous earth is of great commercial value. The oldest commercial *use* is in the form of mild abrasives in silver polishes .and tooth pastes. At one time, it was used as the absorbent for liquid Nitroglycerin in dynamites.

Diatoms are also used in the filteration of liquids especially in sugar refineries, where a small amount of finely powdered diatomaceous earth is added to a sugar solution and the mixture is forced through a filter press. The layer of the diatomaceous earth filters out suspended materials in the liquid.
Diatamaceous earth finds extensive use as an inert mineral filter in various

paints and plastics. In paints, the addition of diatomaceous earth is known to increase the night visibility. It is also an insulator to boilers and blast furnaces where high temperatures are maintained. In temperatures over 1000F, diatomaceous earth is a more effective heat insulator than Magnesium or Asbestos.

Some of the other uses of diatomaceous earth are

(1) It is used as a bacteriological filter.

(2) Cut into small pieces and used as light weight bricks.

(3) Sprinkling of powdered diatomaceous earth on the walls of mines reduces the danger of dust explosions.

Besides these, much of the petroleum of today is believed to be the photosynthetic product of diatoms. Because of the exquisite nature and fineness of the wall sculpturing, they are often used to test the resolving capacity of the microscope. No wonder, the diatoms have been called the "Jewels of the plant world".

Phytogeny

The unicellular nature of diatoms and desmids prompted some to suggest a relationship between the two. But Fritsch (1948), points out that this is merely superficial. Pascher has strongly supported a relationship between Bacillariophyceae, Chrysophyceae and Xanthophyceae and has included all of them inChrysophyta, (see also Smith 1955). The main contention for this is the preponderance of yellow and brown pigments in chromatophores in all the three classes *and* the absence of starch as the photosynthetic product. According to Fristch(1948), this indicates a phylogenetic relationship of some significance. Silica deposits also seem to be common in all three classes (mainly in cysts in Xanthophyceae and Chrysophyceae).

Schutt and others believe that Bacillariophyceae has affinities with Dinophyceae, based mainly on a certain similarity in pigmentation and the products of assimilation. But Fritisch(1948), observes "The fundamental organisation of Dinophyceae is so different that there is no basis for such a view".

10

ACTINOMYCETES

Actinomycetes or 'Ray fungi' (*Acti* = rays; *myces* = fungus), were first described by Harz in 1878, in connection with the microorganisms found in the pus of cattle suffering from the disease – lumpy law. The disease is now called *Actinomycosis.*

Actinomycetes though traditionally classified among bacteria have a number of differences with them. They have a branching mold like body which has the typical mycelial structure seen in fungi. Filaments (mycelia) may be short or long and produce spores just like fungi. Most of our antibiotics are obtained from this group of organisms. Selman. A Walksman was the first one to isolate the first broad spectrum antibioti (streptomycin), from *Streptomyces griseus.*

The following are some of the general characters of Actinomycetes.

1. Majority of them have a branched mycelial body.
2. They are prokaryotic and gram positive.
3. They are not sheathed, stalked or photosynthetic.
4. Most of the Actinomycetes are chemoorganotrophic, although some of them grow on simple mineral media.
5. Cell wall has some peptidoglycan as in gram negative bacteria, but there is variety in the peptidoglycan composition more than in gram negative bacteria.
6. Cells have a diameter 0.5μ to 5μ. In some branched members, the filaments may be as long as several millimetres.
7. The filament is made up of several cells but in some, septa are absent and the members or coenocytic.
8. Some species are motile but most of them are not.
9. Most of them are saprophytic, widely distributed in the organic matter in soil, dung and marine and fresh waters. Some are pathogenic parasites *(e.g. Mycobacterium).*

10. Reproduction is by conidia borne conidiophores. Endospore formation is generally not seen.
11. In certain families, filaments tend to break and fragmentation leads to occoid or elongate cells which develop into new individuals.
12. Except for some (Actinomycetaceae), majority are aerobic.

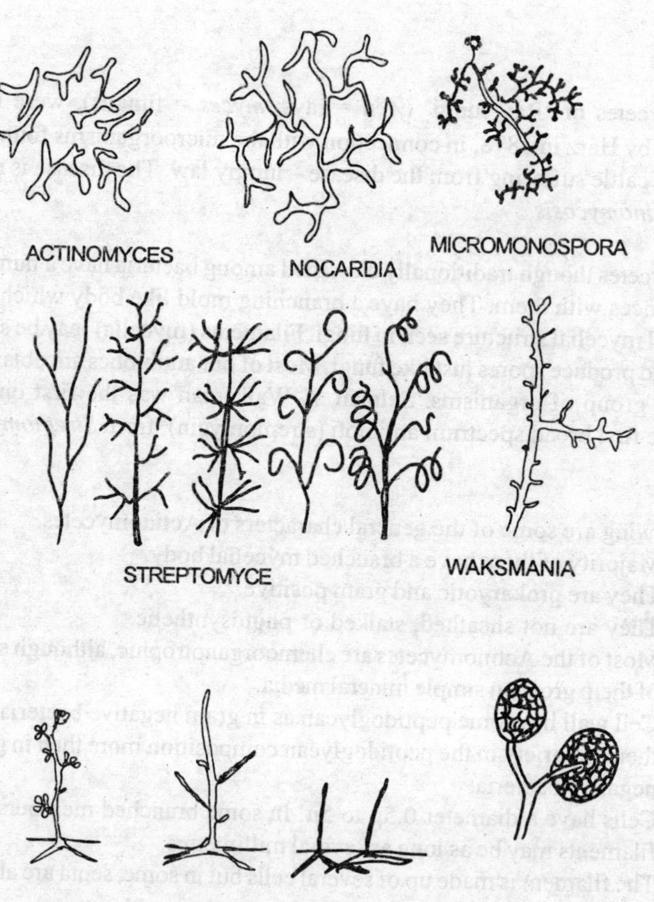

Fig. 10.1 Actinomycetes
Some actinomycetes showing spore bearing hyphae

Difference between bacteria and actinomycetes.

Actinomycetes differ from true bacteria and fungi in the following respects.

1. Actinomycetes have true branching, bacteria do not.
2. Actinomycetes are not sheathed or photosynthetic, while bacteria (atleast some) are.
3. Actinomycetes do not accumulate iron, sulphur or other free elements in or on the cells, while bacteria do.
4. Actinomycetes do not form endospores.

Differences between Actinomycetes and Fungi

1. Actinomycetes are much smaller (1-5 μ in diameter and not more than few m.m in length) than fungi (10-20 μ in diameter; mycelial length varies).
2. Actinomycetes are prokaryotic.
3. Cell walls of Actinomycetes contain mucopolysaccharides and both muramic and diaminopimelic acid as in bacteria, while fungal cell walls are chitinous.
4. Sexual reproduction seen in fungi is absent in Actinomycetes.

Classification

Actinomycetes are traditionally classified as a part of the bacteria. According to Bergey's Manual of Determinative Bacteriology, Actinomycetes are included in several sections of volume IV. It is true however, that Actinomycetes are neither true fungi nor true bacteria. All Actinomycetes are included under the order Actinomycetales.

The order Actinomycetales is divided into four families – *Streptomycetaceae. Actinomycetaceae, Actinoplanaceae* and *Mycobacteriaceae*.

Actinomycetes have also been classified into several groups based on biochemical parameters. Four groups have been identified in Actinomycetes based on the major cell wall constituents.

Cell Wall Type	Sugar Pattern	Genera
I	No Characteristic sugar pattern	*Streptomyces, Streptoverticillicum etc.*
II	Araginose, xylose (monosaccharide)	*Actinoplanes, micromonospora etc.*
III	No Sugars	*Dermatophilus, Planomonospora etc.*
IV	Galactose, Arabinose	*Mycobacterium, Nocardia etc.*

We will now discuss in some detail the characteristic features of the various families of Actinomycetes.

Streptomycetaceae

The family includes two genera *Streptomyces* and *Micromosospora. Thermactinomyces,* a third genus is often included in the family.

According to some microbiologists, the family has four genera. These are *Streptomyces, Streptoverticillum, Sporrichthya,* and *Microcellobospora.*

Streptomyces has long much branched aerial mycelia (mycelia grow above the substratum) which are ceonocytic. The hyphae range in diameter from 0.5 – 2μm. The hyphae do not readily fragment. Cell wall has some chitinous material like in fungi, but it is not true chiten. *Streptomyces* multipliey by means of conidia produced asexually on conidiophores or sporophores. Conidia at the tips of condiophores form long straight or curved chains of spores. The direction and method of branching in the hyphae and conidial coils are of systematic value in delimiting the species in *Streptomyces.*

The hyphae of *Streptomyces* are heterotrophic, aerobic and highly oxidative. In culture conditions, colonies are usually tough, dense textured and adhere to the medium due to the sub surface growth of the mycelium. Many species of *Streptomyces* produce a wide variety of pigments red, orange and yellow. Colonies usually range in diameter from Imm to several milimeters.

Colonies of *Streptomyces* produce a musty damp odour due to the release of a volatile oil *geosmin.*

The genus *Streptomyces,* has 463 species defined in Bergey's manual and many more of uncertain taxonomic position. From the point of view of economic importance, *Streptomyces* perhaps is the most important as it yields many valuable antibiotics.

There are 40 species in the genus *Streptoverticillium.* It produces both sub surface and aerial mycelia. Aerial mycelium appears like a barbed were with short branches. Reproduction is by spores as well as fragmentation. The genus produces a wide variety of antibiotics and pigments.

Sporichthya, another genus of the family has short branched hyphae. Aerial hyphae divide into smooth walled spores which are flagellate.

In the genus *Microcellobospora,* the hyphae are slender and about 1mm in diameter. Substrate mycelium (horizontal mycelium) grows into a compact mass. Sporangia are borne on sporangiophores. Both aerial and substrate hyphae produce sporangiophores.

The genus *Micromonospora,* which is often included in a separate family *Micromonosporaceae* has only substrate mycelia. Aerial hyphae are not conspicuous. Hyphae are coenocytic. Conidia occur singly or in small clusters at the tips of the hyphae.

Actinoplanaceae

Among all the Actinomycetes it is the members of Actinoplanaceae which are most mold like. They are commonly found in soil and acquatic habitats. Occasionally, they also grow on leaf litter as saprophytes. The vegetative mycelium is much branched, sparsely septate with both substrate and aerial hyphae. Diameter of the hyphae varies from 0.2–2.6 mm. A sexual reproductive structures are borne in sporangia under aerobic conditions. Spores are flagellate like in *Saprolegnia* (a fungus belonging to the class phycomycetes). The family has 10 genera which fall into two distinct groups – (i) members with large, spherical or irregular sporangia producing many spores and (ii) members with small filiform club shaped sporangia containing one of the few spores.

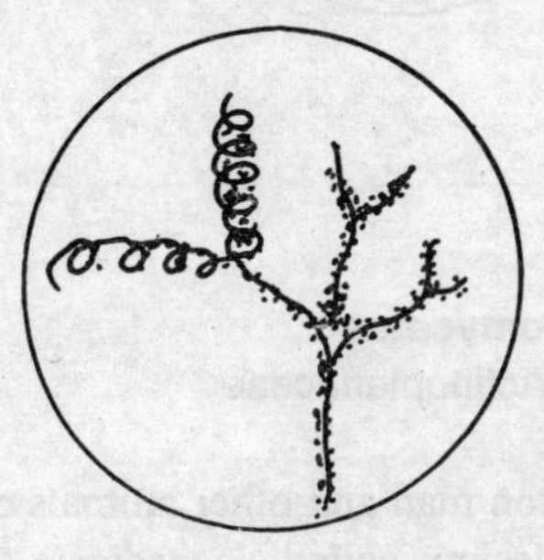

Fig. 10.2 Actinomycetes
Filament formation in Streptomyces

Actinomycetaceae

The family includes two genera – *Nocardia* and *Actinomyces*. Both the general produce branched mycelia. *Nocardia* is aerobic, while *Actinomyces* is a strict anaerobe.

Nocardia is similar to *Streptomyces* in structure, but differs from the latter in (i) the filaments are ceonocytic, later become septate and tend to fragment readily into bacillus like or coccoid segments and (ii) no conidia or spores are formed. *Nocardia* is a saprophyte that lives in soil as decomposing agent of a variety of substances such as cellulose, proteins, lipids and polysaccharides. The mycelium in cultures grow as well as in the medium.

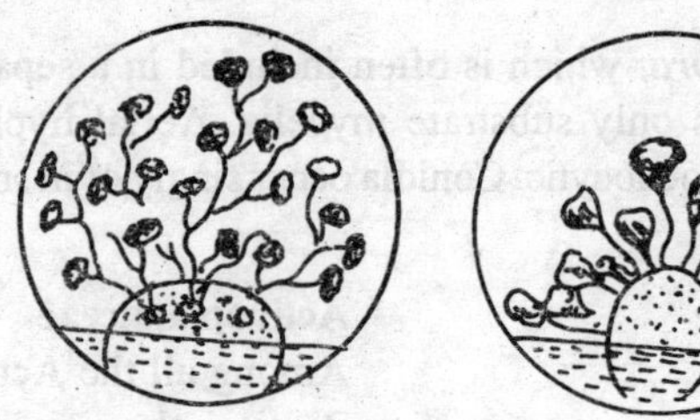

Fig. 10.3 Actinomycetes
Fruiting structures in Actinoplanaceae

Some species of *Nocardia* are parasitic on man and other animals causing tuberculosis type diseases, or ulcer like lesions called *Nocardiosis*.

Pseudonocardia is a closely allied genus of *Nocardia*. It differs from the latter in having spores borne on specially differentiated hyphae.

The genus *Actinomyces* is an obligate anaerobe requiring complex organic food (like chopped meat). In culture both *Actinomyces* has morphology similar to *Nocardia.* Several species are pathogenic to human beings and animals.

Actinomyces bovis which causes the *lump jaw* disease in cattle, occurs normally in the oral cavity of man and cattle. If by an injury it (*Actinomyces*) is introduced into the flesh of tongue or cheek it produces swellings. This results in what is popularly called 'wooden tongue' in cattle. Microscopic examination of the infected tissue reveals that the mycelium grows in compact granulara mass acquiring a bright yellow colour. These are called 'sulphur granules' and contain radiating mycelia held in a hard matrix of a polysaccharide protein complex that is also rich in calcium phosphate.

Mycobacteriaceae

The family includes two genera *Mycobacterium* and *Micrococcus.* The genus *Mycobacterium* has over 30 species that include soil saprophytes and two human pathogens – *M.tuberculosis* and *M.leprae* causing tuberculosis and leprosy respectively. The *Mycobacteria* are mostly bacteria like and rarely form true mycelia. Cells are slightly curved or straight rods, sometimes branching. Short filaments that may be formed tend to break up soon into bacillus like fragments.

A unique staining property distinguishes *Mycobacterium* from other bacteria. Called *acid fastness,* this is associated with the waxy components in the cell walls. *Mycobacteria* are gram positive, non-motile, aerobic and do not form spores. They can be cultured on rich, solid, organic media.*M.Tuberculosis* the human tuberculosis causing pathogen, was first isolated by Robert Koch in 1882. The tuberculosis bacilli can live outside the host in dried sputum for as long as 6–8 months. They get distributed in the air through the dried sputum and may get inhaled. Exposure to bright sunlight and pasteurization kills them. There are several kinds of tubercle bacilli varying according to the host. These are :

1. *Mycobacterium tuberculosis var hominus* (Human beings)
2. *M. tuberculosis* var *bovis* (Cattle)
3. *M. tuberculosis* var *vole* (field mouse)
4. *M. tuberculosis* var *avium* (birds)
5. *M. tuberculosis* var *marinum* (fish and other cold blooded animals)

The BCG bacillus used for vaccination is a modified bovine strain.

Life cycle of Actinomycetes

Some microbiologists believe that there are two phases in the life cycle. A haploid substratal mycelium called the primary mycelium, and a diploid aerial secondary mycelium. Electron microscopic studies of *Streptomyces griseus* has shown the presence of certain more or less spherical structures believed to be nuclei. Both nuclear fusion and reduction division are believed to take place before vegetative growth sets in.

Actinophages

Viruses attacking Actinomycetes are called actinophages. They have been isolated from Streptomyces and are specific to it. Actinophages are similar to bacteriophages in most respects. It is for this reason too, Actinomycetes are believed to be closer to bacteria than fungi.

Genetics of Actinomycetes

General studies on Actinomycetes have revealed that frequent variations occur in them, in cultures, much the same way as in bacteria These variations can be divided into three categories.

(i) Adaptive variations
(ii) Continuous or fluctuating variations and
(iii) Development variations in saltants or mutants

Many microbiologists have reported morphological variations with R (Gods) and F (C filaments) forms of filaments. Variations (due to sub culturing), like change of pigmentation, non sporulation, loss of synthetic powers of certain amino acids have been commonly found in several Actinomycetes. The spontaneous occurrence of saltants and mutants has been observed in *Streptomyces.* Mutations have been induced in many Actinomycetes with the help of UV, Xrays and chemicals. Fortified strains of *Streptomyces* with reference to the production of antibiotics have been obtained by exposing them to the combined action of UV and chemicals. Drug resistance has also been reported in many Actinomycetes. *Streptomyces coelicolor. S.griseus* and *S.fradiae* have exhibited phase resistance in culture media, and phage resistance colonies have been obtained.

Genetic recombination has been demonstrated in *Streptomyces griseus* using nutritional and drug resistance genes as markers. The resultant hyphae combining both the parental characters are believed to be fusion products.

The spores produced by the off spring, however, belong to only one parental type indicating that the recombinants are not stable. The recombinants can multiply only by fragmentation but not by spore formation.

Economic importance

Actinomycetes are economically important as they are both harmful and useful. It is noteworthy that while some Actinomycetes cause diseases, the remedy also comes from the same group.

Many of the Actinomycetes cause human and animal diseases. Actinomycosis, Avian tuberculosis, human tuberculosis are all caused by Actinomycetes.

Species of *Actinomyces* cause a number of diseases. *A. israeli* produces granulomatous swelling of jaw, lungs, abdomen etc. The organism spreads from the mouth. *A bovis* is a parasite causing actinomycosis in cattle, lung infections in rabbits, and guinea pigs. *A.madurae* cause granulomatous disease of foot in man known as Madura foot.

Nocardiasteroides causes chronic granulomatous infections in intestines and lungs as well as abscesses of the feet.

Streptomycin, one of the greatest of antibiotics discovered by Walksman, is a great weapon in the fast against tuberculosis. It is obtained from the mold *S.griseus,* while tuberculosis is caused by its cousing *Mycobacterium tuberculosis.* Many other antibiotics like chloromycetin (*S.venezulae*),

Terramycin (*S.vinezulae*), Aureomycin (*S. aureofaciens*), Rifampicin (*S.mediterranei*),. Nystatin an antifungal agent (*S.noursei*), Bleomycin (*S.verticullus*). Viomycin (*S.floridae*), Vanomcin (*S.crientalis*), Novobiocin (*S.niveus*), Cycloserine (*S.archidaceus*), Lincomycin (*S.lincolnesis*), Natamycin (*S.natalensis*), Mitomycin (*S.caepitosus*), Puromycin (*S.alboniger*), and Dactinomycin (*S.antibioticus*) etc. are produced by Actinomycetes –specially by *Streptomyces.*

FUNGI

FUNGI represent one of the most important groups of the plant kingdon. The study of Fungi is known as *Mycology*. Etymologically, mycology means the study of mushrooms [Gr. *Mykes* = mushroom *logos* = study]. Though the name mycology was originally meant to include only the mushrooms, now a days it has a broad meaning. It encompasses all the fungi. The reason for mushrooms being the first to be studied among fungi is that they are one of the largest among fungi, and hence attracted the attention of naturalists even before the invention of microscope.

With Leeuwenhock inventing the microscope in the seventeenth century, the systematic study of fungi began. Pier Antonio Micheli–the Italian botanist did a great lot to found the science of mycology, and with justice he deserves to be called the Father of mycology. In 1729, he published *Nova plantarum genera,* in which he included his researches on Fungi.

What are Fungi?

It is rather difficult to define the term fungi giving enough justice to its scope. Without entering into any hair splitting argument, we can generally accept the definition given by Alexopoulos (1952). He defines fungi as "*nucleated spore bearing achlorophyllous organisms which generally reproduce sexually and asexually, and whose filamentous, branched, somatic structures are typically surrounded by cell walls containing cellulose or chitin or both*"".

Present day Fungi include such common plants like mushrooms, puffballs, toadstools, morels, rusts, smuts and a host of other members.

Fungi and Mankind :

Of what importance is the study of fungi to mankind? This question can be very easily answered by listing both the good and the bad done by fungi to man.

Fungi are friends as well as foes of mankind. Though the systematic study of fungi is only two and a half centuries old, the manifestations of this great group of plants on human beings has been felt since time immemorial. Alexopulos, in a lighter vein mentions "Ever since the first toast was proposed over a shell full of wine, and the first loaf of bread was baked, fungi are known to mankind". It is perhaps impossible to imagine how the world would have been without these fungi. Without fungi, of course we would not have had such deadly plant diseases like rusts and smuts which take such a heavy toll of our wheat crop. Our potatoes would not have had the notorious late blight, areca palms would have been safe without kole roga, grapes would not have suffered with mildew disease, and we need not have bothered about our food articles which now get the attack of molds; not much of care would have been needed to maintain aseptic conditions in the laboratory. Our clothes, shoes etc. would be free from the green molds and we would not have heard such deadly diseases like monilasis, ringworm etc. Well, one may exclaim the world would have been a real paradise to live in without these ugly creatures. But, at the same time let us view the other side of the coin. Mankind would not have heard anything about alcoholic beverages–the famous champagne would be non-existent, there would be no bakeries and perhaps there would be dead bodies of organisms allaround–for there would not be any saprophytic fungi to decompose them, there would not have been that great potent drug–*penicillin*–the cinderella of antibiotics, and we would not have had such fine delicacies like mushrooms and morels; well these are a few of the favourite things we would have missed. And really, we cannot afford to miss these. To provide both good and bad, fungi are there, and it is in his own interest that man is striving to have an intimate knowledge of these organisms, so that he can improve the races in which he is interested and try to check the growth of others that are harmful to him. The above explanation is just a representative of what the fungi are capable of doing.

General Characters

Fungi comprise more than 3,900 genera and 80,000 species.

Occurrence

Fungi have a great potentiality for quick adaption, hence, they have colonised all possible types of habitats. Primitive fungi are aquatic in their habitat. Gradually higher and higher fungi adopt amphibious and terrestrial habitats. Majority of the terrestrial forms are saprophytic in the soil, or on animal and plant remains.

The fungi constitute a group of living organisms, which do not have chlorophyll roots stems or leaves that are familiarly met with common plants. They are usually filamentous and multicellular (could be unicellular also). With few exceptions, they exhibit little division of labour.

Nutrition and growth

As fungi are achlorophyllous, they obtain nutrition by heterotrophic mode. This is of 2 types.

(1) *Parasitic method.* (Gr. *Parasitos* eating beside another) Parasitic fungi live on other organisms (plants, animals including human beings), and obtain their nutrition from a living host. In the process, they may cause disease to the host. Parasites may be ecto-parasites or endoparasites depending on their location in the body of the host. Ectoparasites live on the body of the host (Ex Erysiphales), while endoparasites live inside the tissues of the host.

Parasites may also be classified into obligate parasites and facultative parasites. Obligate parasites are those which can live only on a living protoplasm, while facultative parasites depending on the circumstances, may live on dead organic matter or on living organisms.

(2) *Saprophytic method* (Gr. *Sapors* = rotten + *bios* = life). Fungi which obtain nutrition from dead and decaying organic remains are called saprophytes, Saprophytic fungi can be easily cultured in the laboratory. Saprophytes are also of 2 types. (1) *Obligate saprobes* and (2 *Facultative saprobes.* Fungi which live on dead organic mater and are incapable of infecting living organisms are called Obligate saprobes; those capable of causing disease on a living host of living or dead organic matter depending on the situation are called Facultative saprobes.

Mycorrhiza

Some fungi enter into an association with the roots of higher plants and it is known as mycorrhiza (Gr. *Mykes* = mushroom +*rhiza* = root).

In many cases the association is mutually beneficial. Though much has been discovered (Kelly 1950) with regard to mycorrhiza, the exact relationship is doubtful (Alexopoulous 1952).

Fungi can synthesise all their nutritional requirements if they are initially supplied with an organic food such as carbohydrates. Laboratory studies have indicated that fungi require all the essential and trace mineral elements for their growth. They can synthesise their vitamin requirements generally.

While some of the fungi are *omnivorous* and can grow on any moist organic matter, others are specific about nutrition, and grow only on particular species. Most fungi are capable of growing in the temperature range between 0° and 35°C, the optimum lying between 20°–30°C.

But they can withstand at least for a few hours such low temperatures as 195°C. As a contrast to bacteria, fungi prefer an acidic medium for growth, with a pH of 6 being most suitable.

While light is not very much necessary for the growth, it is needed at least at the time of spore formation.

Somatic structure

The plant body of a fungus generally consists of thread or tube like cylindrical filamentous structure called *Hypha* (pl. *hyphae,* Gr. *Hypha* = web). A hypha consists of a thin wall and has an inner lining of cytoplasm. A hypha may or may not have cross walls. The cross walls are called septa (sing. L. *septum* = hedge of partition). A hypha with septa is called septate; without, it is known as aseptate.

Cell wall composition

The chemical composition varies in different fungi. While cellulose is the chief component in some, in most of the higher fungi the wall is mainly composed of chitin. Besides these, hemicellulose, lignin like substances have also been found in fungal cell wall. However, none of these occur in isolation, the fungal cell wall being very complex. The composition of the wall may vary in young and old hyphae, in different pH etc.

All fungi have a definite nucleus. During nuclear division, definite chromosomes are organised. Some mycologists believe that nuclear division in fungi is not directly comparable to mitosis.

The aseptate hyphae usually have a number of nuclei distributed in them, the condition being termed *coenocytic.* (Gr. *Koinos* = common; *Kytos* = a hollow vessel). In septate hyphae the cells may be uninucleate, binucleate or multinucleate. Vacuoles, oil drops and other inclusions are generally present in the cytoplasm.

A fungi plant body usually consists of a mass of hyphae called *mycelium* (pl. mycelia). In forming a plant body, the mycelia may be lose or form thick strands called – *Rhizomorphs* (Gr. rhizos = root + morphal = shape). In this, the mycelial strands are closely united to form root like structures exhibiting division of labour. Rhizomorphs are resistant to adverse environmental conditions.

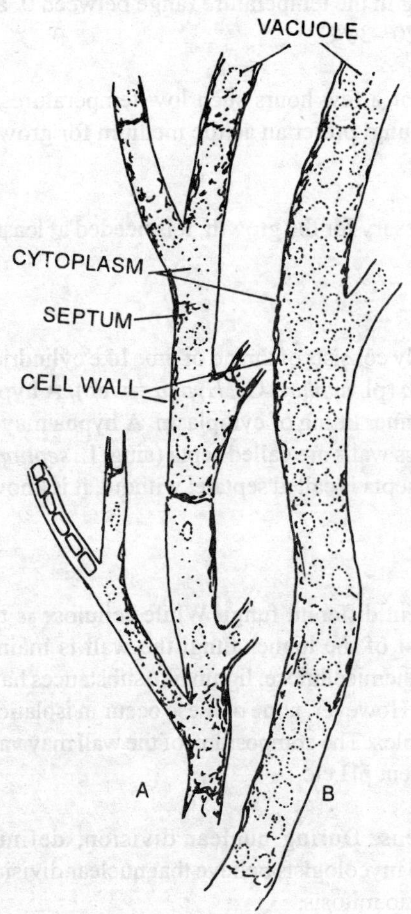

Fig. 11.1 Fungi
Somatic Structures of Fungi
A. Septate hypha,
B. A septate and coenocytic hypha

In parasitic fungi, the hyphae produce special absorbing branches to draw nutrition from the host. These are called *haustoria* (sing. *Haustorium. L.haustor* = drinker). Haustoria are produced in mycelia which are intercellular. They may be knob like, elongated or branched like a miniature root system (Fig. 15.2B). Haustoria are not produced in artificial culture.

Saprophytic fungi do not produce haustoria as they absorb the food by direct diffusion causing disintegration of the organic matter.

Fungal hyphae have a great longevity. As Alexopoulos states, fungal colonies of more than 400 years old are known in nature and it is probable that some mycelia are thousands of years old.

A mycelium starts its life as a short germ tube produced by the germination of a spore (Gr. *Spore* = Seed). Spores are asexual propagating units of fungi.

In higher fungi, the mycelium has a tendency to organise into compact, or loosely interwoven tissue to form compact macroscopic structures. The, general term Plectenchyma (Gr. *pleko* = I weave + *enchyma* = infusion, i.e. a woven tissue) is used to designate all types of fungal tissues. There are 2 general types in plectenchyma – Prosenchyma (Gr. *Pros* = toward + *enchyma*

=infusion i.e., approaching a tissue), is a loosely interwoven tissue in which they hyphae can be identified. They are arranged more or less parallel to one another. *Pseudo-parenchyma* (Gr.*pseudo* = false + *parenchyma* = a type of plant tissue), consists of closely packed more or less isodiametric or oval cells resembling the parenchyma tissue of higher plants. In this type of tissue, the hyphae lose their identity and are not distinguishable.

Prosenchyma and pseudoparenchyma tissues compose many somatic structures in higher fungi. Two such structures are *Stroma* (pl *stromata,* Gr. *Stroma* =mattress), and the *Sclerotium* (pl. *sclerotia;* Gr. *skleros* = hard). A stroma is a flat compact matter like structure in which fructifications are formed. A sclerotium is a hard resting body which is resistant to adverse environmental conditions. It remains dormant and germinates during favourable seasons. (Fig. 15.4).

Reproduction

Fungi reproduce both asexually and sexually. In some fungi, during the formation of reproductive structures (asexual or sexual), the entire thallus is used up. Such a case is common in lower fungi and they are called *Holocarpic.* (Gr. *Holos* = whole + *Karpos* = fruit). In higher fungi, however, only a portion of the thallus produces reproductive structures so that vegetative and reproductive phases can occur together. These fungi are called *Eucarpic* (Gr. *Eu* – good + *karpos* = fruit). Holocarpic nature is regarded as primitive.

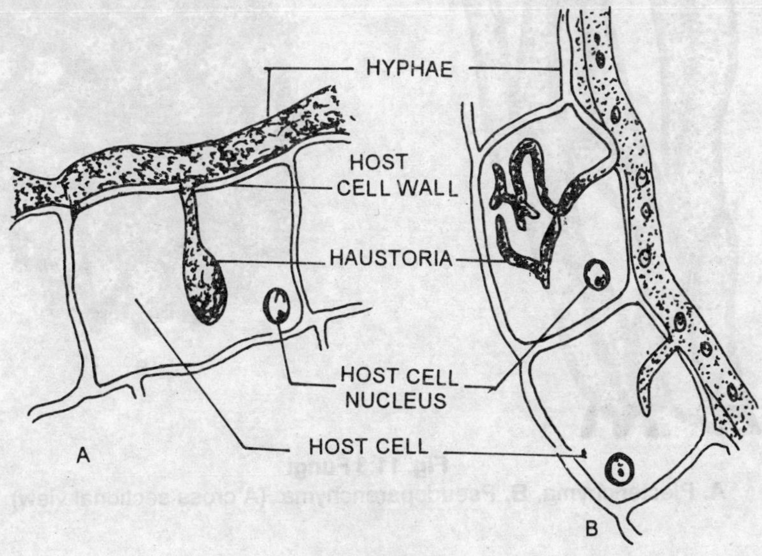

Fig. 11.2 Fungi

Types of Haustoria in Fungi, **A.** Saccate haustorium, **B.** Branched haustoria (Hyphae are inter cellular)

Asexual Reproduction

Asexual reproduction is of frequent occurrence in fungi than sexual, and it is that which accounts for the rapid distribution of fungi. Asexual reproduction takes place by a variety of methods. They are: (1) Fragmentation, (2) Fission, (3) Budding, and (4) Spore Formation.

Fragmentation

This is of 2 types; (a) accidental and (b) purposive. In the first type, when the mycelium accidentally breaks up into several fragments each one will develop into a new individual. In the second type, fragmentation of the mycelium takes place as a normal means of propagation. The hypha breaks up into individual cells. Each such isolated cell is called *Oidium* (pl. *Oidia,* Gr.*Oidion* = small egg), or Arthrospore (Gr. *arthros* = joint + *spore* = seed). Each arthrospore develops into a new individual. During unfavourable conditions arthrospores surround themselves with a thick wall; such spores are *Chlamdospores* (Gr.Chlamys =mantle). They can withstand the unfavourable conditions.

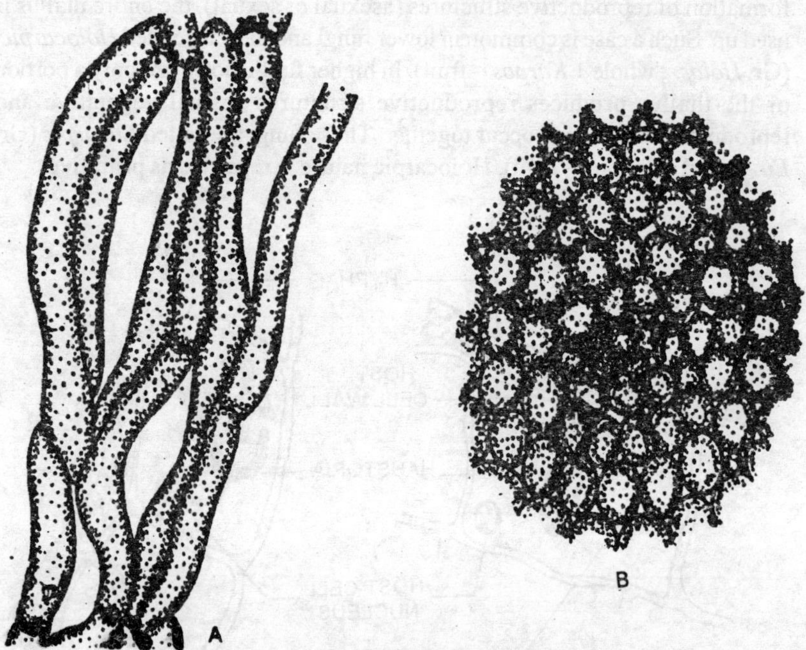

Fig. 11.3 Fungi
A. Plectenchyma, **B.** Pseudoparenchyma (A cross sectional view)

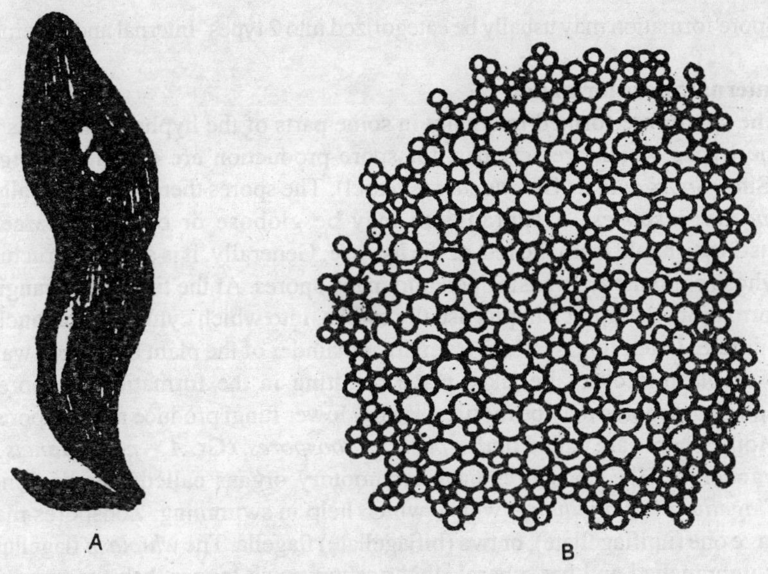

Fig. 11.4 Fungi
A. Entire sclerotium, B. Cross sectional view of sclerotium

Fission

It is the simple splitting of the cell into two daughter cells. First, construction of the nucleus and of the cell takes place, followed by the division of the nucleus and formation of a cell wall. Fission as a means of reproduction is seen in unicellular forms like yeasts and bacteria.

Budding

The production of new cell by the outgrowth *(bud)* of a cell is called budding. As the bud is being formed, the nucleus in the parental cell divides with one of the daughter nuclei entering the bud. The bud eventually breaks from the parent cell, and develops into a new individual.

Spore formation

By far, the commonest method of sexual reproduction is spore formation. Spores vary in colour from hyaline to green, yellow, orange, red, brown to black; in size, from minute to large, in shape from globose to oval, oblong, needle shaped to helical; in number of cells, from one to many; in the arrangement of cell and in the way in which the spores themselves are borne (Alexopoulos 1952). As Alexopoulos states "it is this infinite variety of spores that makes the study fascinating". While some fungi produce only one type, others produce many types of spores.

Spore formation may usually be categorized into 2 types–Internal and External.

Internal Spore formation

The spores are formed internally in some parts of the hyphae. The parts of the hypae which are set apart for spore production are called *sporangia* (Sing. *Sporganium;* Gr. *angeion* = vessel). The spores themselves are called *sporangiospores.* A sporangium may be globose or oval, or scarcely distinguishable from the rest of the hyphae. Generally, it is sac like structure whose contents are transformed into many spores. At the time of sporangial formation, the tip of a hypha usually swells, into which cytoplsm and nuclei migrate. Later, this is cut off from the remainder of the plant by a cross wall. Protoplasmic cleavage takes place resulting in the formation of spores. Sporangiospores may be motile. Usually lower fungi produce motile spores. Motile spores are called *planospores* or *zoospores.* (Gr. *A = not +planets =* wanderer). The zoospores have locomotory organs called flagells (sing. *Flagellum* L. *flagellum* = whip), which help in swimming. Zoospores may have one (uniflagellate), or two (biflagellate) flagella. The *whiplash* flagellum is unbranched and has a basal rigid portion much longer than the terminal flexible portion. The *tinse* flagellum is a branched feathery structure consisting of a rachis with lateral hair like projections.

Flagellar Apparatus

This is a complex structure consisting of (a) *flagella* (b) *blepharoplast,* and (c) *rhizoplast.* Blepharoplast (Gr. *blepharis* = eyelash), which is the base of the flagellum is inside the cell. Rhizoplast is a thread like structure through which the flagellar base is connected to the nucleus. Electron micrographs have revealed that the flagellum consists of eleven fibres of which nine are peripheral and two central. This 9+2 arrangement is seen in all motile cells except in bacteria.

The zoospores are usually pyriform in shape. In some cases, they are reniform also. The flagella are terminal in pyriform and in lateralreniform zoospores. On liberation from the sporangium, zoospores directly germinate to give rise to a new additional.

External Spore formation

In some cases, the spores are formed by the cutting off of the tips of hypae in basipetalous succession. There are specialised branches of the hyhpae for this purpose. They are called *conidiophores* (Gr. *konis* = dust, *phore* = stalk). The spores themselves are called conidia (sing. Conidium). Conidia are non-motile. They are wind dispersed. Either they directly germinate and give rise to a new individual, or produce zoospores which in turn develop into new individuals. Conidia may be uninucleate or multinucleate.

Sexual Reproduction

As in other organisms, the ultimate aim of sexual reproduction is the genetic recombination which is accomplished by the fusion of compatible nuclei.

Fungi may be *homothallic* or *heterothallic.* In the former, both sex organs are formed on the same thallus, and in the latter they are formed on different thalli.

The sex cells are gamets. They are produced in sex organs. Sex organs are called *gametangia,* if they produce gamets internally, or *Gametophores,* if they produce gamets externally. Gametophores, however are found only in the case of male. (Gr. *Gametos* = husband +*angeion* = vessel). If gametangia produce only one type of gamets they are called *isogametangia,* and the gamets themselves are called *isogametes.* If the gametes are of two different types, such gametes are *anisogametes* or *heterogametes,* and the gametangia are called heterogametangia. *Heterogametangia* are two types of male and female.The male gametangiium is called the antheridium, and the female oogonium (Gr.*antheros* = flowery, *oon* = egg + *gonos* = off spring). (The female sex organs in Ascomycetes is *ascogonium).*

Having thus explained the nature of sex organs, we may now learn the different types of sexual reproduction met with in fungi. They are :

(1) Planogametic Copulation

This type is mostly seen in lower fungi. This involves the fusion of two motile gametes (planogametes). The gametes may be *isogametes* or *anisogametes.* In this case, fusion is outside the gametangia. This is seen in some aquatic fungi belonging to lower *phycomycetes.*

(2) Gaetangial Contact

In a large majority of fungi, the male and female gametes are reduced to undifferentiated uninucleate protoplasts. Here, the sex organs are *antheridia* and *oogonia.* In this method, the two sex organs come in contact. Usually to effect the contact, a *conjugation tube* or *fertilization tube* is formed from the antheridium which penetrates the oogonium.. This forms a passage for the migration of male nuclei into the oogonium. After the nuclei migrate, the antheridium disintegrates.

(3) Gametangial Copulation

In this method, the entire contents of the contacting gametangia fuse. At the time of reproduction, the two gametangia come nearer and become closely appressed. The walls in between the two dissolve and the antheridial contents flow into the oogonium.

(4) Spermatization

In some higher fungi, special erect branches of the hyphae (*Spermatiophore*) are produced which cut off uninucleate male reproductive cells called *spermatia.* (Gr. *spermation* = little seed). The spermatia are formed much the same way in which conidia are formed. The spermatia are carried by wind or insects to the female reproductive structure – the *Receptive hypha,* the basal cell of which is fertile. The spermatium settles at the tip of the receptive hypha. The walls in between the two dissolve and the spermatial nucleus migrates into the receptive hypha.

(5) Somatogamy

This is the most advanced type of reproduction in fungi. In this type, no sex organs are formed. At the time of reproduction, two vegetative hyphae of compatible mycelia, fuse. This is met with in higher ascomycetes and higher basidiomycetes.

Stages in sexual reproduction

Generally, in lower fungi like phycomycetes, sexual reproduction is accomplished in three phases. In these forms, haploid phase is predominant with a short diploid phase restricted to the zygote. The three phases are :

(1) *Plasmogamy (Gr. plasm* = a moulded object + *gamos* = marriage, union). In this, the protoplasts (male and female) fuse bringing the nuclei closer.

(2) *Karyogamy, (*Gr. ***Karyan*** = nut, nucleus). In this phase, the two nuclei fuse.

(3) *Meiosis* (Gr. *meiosis* = reduction). This immediately follows karyogamy and restores the haploid complement of the chromosomes. In short, plasmogamy fuses the two protoplasts. Karyogamy brings about the fusion of two nuclei resulting in the formation of a diploid nucleus. Meiosis restores the haploid condition.

In higher fungi (Ascomycetes and Basidiomycetes) however, after *plasmogamy, Karyogamy* is delayed for a very long time. In the intervening phase, the male and female nuclei just remain together without fusing. This condition is called the *Dikaryotic phase.* Only at the end of the life cycle, karyogamy takes place which is soon followed by meiosis.

As a result of the dikaryotic phase in the life history, three types of mycelia (with reference to nuclar complement) alternate. They are :

(1) *Primary mycelium.* This is *monkaryotic* and *haploid.* Reproducing sexually after *plasmogamy,* this produces

(2) *Secondary mycelium* which is dikaryotic. This occupies much of the life cycle. At the end of the life cycle, in some parts of secondary mycelium karyogamy take place. These are

(3) *tertiary mycelia* which are diploid. *Karyogamy* is soon followed by *meiosis* resulting in the formation of haploid cells, which germinate to give rise to primary mycelium.

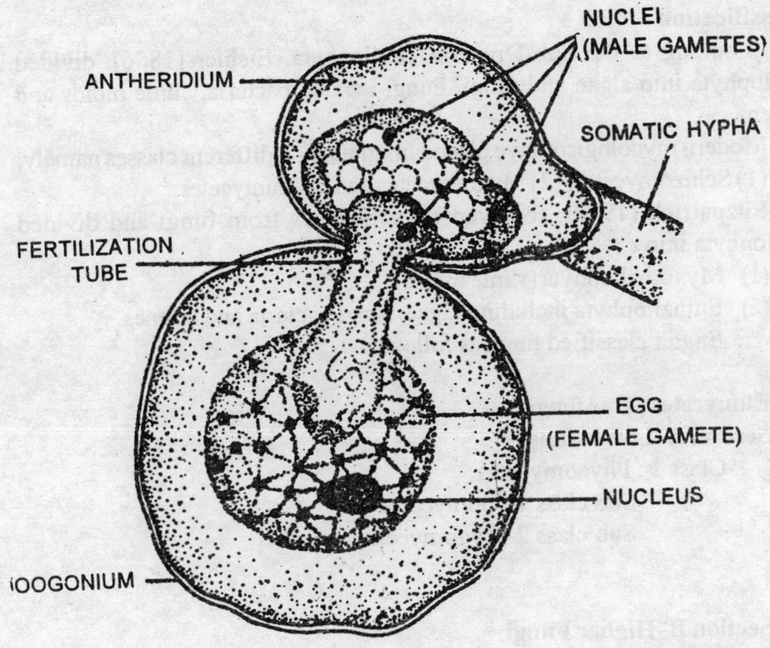

Fig. 11.5 Fungi
Antheridium and oogonium showing Gametangial contact

Sexuality

As we go from lower to higher fungi, there is a gradual elimination of sex organs and in higher fungi sex organs are entirely wanting. Sexually reproduction here takes place only by *somatogamy.* But this should not be taken as a loss of sexuality. Because in somatogamy there is no random fusion of any two hyphae. The *compatible hyphae* are carefully selected, the compatibility being determined by genetic facrtors. Many of the higher fungi though lack sex organs are, heterothallic i.e., they are self incompatible.

Heterothallism is of two types.

(1) ***Bipolar heterothallism*** in which sexuality is governed by a set of factors (AB), and *tetrapolar heterothallism,* in which sexuality is determined by two sets of factors (AaBb). In these two cases, only thalli which carry the opposite factors (a and B in bipolar, AB and ab in tetrapolar) can mate.

This clearly indicates that there is no loss of sexuality in higher fungi. In fact it is the most sophisticated type of sexual differentiation.

Classification :

Fungi belong to the Sub-Division Thallophyta. Eichler (1886), divided thallophyta into algae and fungi. Fungi include bacteria, slime molds and lichens.

Modern mycologists have placed fungi under 3 different classes namely; (1) Schizomycetes, (2) Myxomycetes and (3) Eumycetes.

Fitzpatrick (1930), removed myxomycetes from fungi and divided thallophyta into :

(1) Myxothallophyta (slime molds)
(2) Euthallophyta including algae, fungi bacteria and lichnes.

Engler classified fungi as follows;

Eumycetes (True fungi)

Section A–Lower fungi

Class 1 : Phycomycetes.

Sub class 1–Oomycetes

Sub class 2–Zygomycetes.

Section B–Higher Fungi

Class 2 : **Ascomycetes**

Sub class 1 – Hemiasci

Sub class 2 – Euasci

Class 3: **Basidiomycetes**

Sub class 1 – Hemibasidii

Sub class 2 – Eubasidii

Section C– Deuteromycetes or Fungi imperfectii.

Clement and Shear (1931) divided fungi into 5 divisions –
(1) Phycomycetes, (2) Ascomycetes, (3) Basidiomycetes, (4) Promyceteand (5) Deuteromycetes.

Gwynne vaughan and Barnes (1926), classified fungi into 5 classes namely,

Phycomycetes, Archemycetes, Ascomycetes, Basidimycetes and Deuteromycetes.

They divided – Phycomycetes into Oomycetes and Zygomycetes;
Ascomycetes into Plecomycetes, Pyre mycetes and Discomycetes
Basidiomycetes into Protobasidiomycetes and Autobasidiomycetes.

Wolf and Wolf (1948), classified fungi into phycomycetes, Ascomycetes, Basidiomycetes, Myxomycetes and Deuteromycetes.

Gauamann and Dodge (1928), followed the classification of Gwynee Vaughan in the main divisions. Tippu (1942), divided Fungi into:

(1) Schizomycophyta Eg. Bacteria
(2) Myxomycophyta Eg. Slime molds
(3) Eumycophyta

This includes four classes :

(i) Phycomycetes (ii) Ascomycetes
(iii) Basidiomycetes (iv) Dweuteromycetes.

Alexopoulos (1956) places all Fungi in the Division Mycota. He, however, keeps away the order Acrasiales and Labry inthilales from Fungi as their affinities are uncertain.

The division mycota is divided into two sub-divisions.
(1) Myxomycotina and (2) Eumycotina
Myxomycotina has only one class – Myxomycetes.
Eumycotina has the following classes;

(a) Chytridiomycetes (b) Hypochytridiomycetes
(c) Oomycetes (d) Plasmodiophoromycetes
(e) Zygomycetes (f) Trichomycetes
(g) Ascomycetes (h) Basidiomycetes
(i) Deuteromycetes.

In this book the following members are discussed.

1. Phycomycetes – *Pythium* and *Rhizopus*
2. Ascomycetes – Yeast, *Claviceps, Aspergillus* and *Penicillium*
3. Basidiomycetes – *Sphacelotheca*
4. Deuteromycetes – *Fusarium*

ORDER – PERONOSPORALES

Family – Phythiaceae

Members of this family are both aquatic and terrestrial. Most of them are saprophytes or facultative parasites living in the soil. Asexual reproduction takes place by means of sporangia, which are borne at the tips of the hyphae.

The sporangia may remain attached to the parent hypha and produce zoospores. Detached sporangia are dispersed by wind or water. The sporangia bearing hyphae may or may not be distinct form the somatic hyphae. Sexual reproduction is *oogamous.* Sex organs are borne terminally.

The two important genera of the family are : *Pythium* and *Phytophthora.*

The genus *Pythium* includes about 66 species [Middleton 1943]. Butler and Bisby (1958), have recorded as many as 19 species from India. Many are amphibious in their habitat. They mostly occur is moist humus soil and infect the hypocotyl region of the seedlings of plants like ginger, tobacco, mustard, wheat etc. After entering the host, they act as parasites and cause diseases like *"damping off", soft rot", "foot rot",* and *"wheat rot",* of seedlings. For a certain period of time no sign of their presence is noticeable on the host, but later, seedlings become pale-green and show a shrivelled girdle of brown colour near the surface of the soil in the region of hypocotyl extending upwards and downwards from the ground level. Due to this, the seedlings collapse, and bend over in a characteristic manner. Hence, the disease is called *"damping off"* of seedlings. Best known example of "damping off" is that of tobacco seedling caused, by *"pythium debaryanum".* Other species causing dissease are: *P. aphanidermatum,* causes "foot-rot" of wheat seedlings commonly found in U.S.A., and Canada, *P.myriotylum* causes *"stem rot"* of ginger in India.

Mycelium

Mycelium is non-septate, branched and septa are formed only during the formation of reproductive structures. Hyphae contain a vacuolated mass of protoplasm in which a large number of nuclei lie embedded, resembling superfically the filaments of *Vaucheria.* Reserve food is the form of glycogen. The infection may occur through the stomata or the germ tubes may pierce through the cell-wall by enzymatic action. Haustoria are not produced by this fungus.

Reproduction

Reproduction takes place both asexually as well as sexually.

Asexual reproduction

Asexual reproduction takes place by the formation of sporangia which are small, globular or sac like structures produced on the aerial hyphae. Sporangia may also be intercalary in position.

Under moist conditions sporangia remain attached to the hyphae. They function as zoosporangia. At maturity, a beak like outgrowth is formed either at the tip or from the side of the zoosporangium. Later, this beak like structure enlarges at its distal end into a bladder like vesicle. The contents of zoosporangium by now would have divided into uninucleate daughter protoplasts. These migrate into the vesicle. Subsequent maturation of zoospores takes place in the vesicle. Zoospores are liberated to the outside by the rupture of the vesicle.

Fig. 11.6 Fungi
Phythium. Damping off of sedling (Note the rooted stem at the base)

Liberated zoospores are reniform with two flagella inserted laterally. One of these is of the tinsel type and the other of the whiplash type. After a short period of motility, they (zoospores) lose their flagella, roundoff and secrete a delicate wall around. In this condition they usually germinate by producing a short germ tube. Hohsik (1932), reports that the contents of the encysted zoopore may form a single zoospore. This emerges from the cyst and germinates to give rise to a fresh mycelium.

In *Pythium,* the asexual reproductive structures namely the sporangia are of two principal types (a) *spheroidal* or *globose,* and (b) *elongate* or *filamentous* (hypal).

(a) Spheroidal or globose sporangium are commonly seen in *P.debaryanmum.* Here, the sporangia are more or less spherical and are borne terminally.

(b) Filamentous or elongate sporangium is indistinguishable from the somatic hyphae. This may be simple as in *P.gracile* and *P. monospermum*, or branched as in *P. apanidermatium* and *P. myriotylum*.

Evolution of Conidium

One of the main evolutionary tendencies among the phycomycetes is the transformation of zoösporangium into a conidium. This is necessary because of the migration from aquatic to sub aerial habitat. Different species of *Pythium* very well illustrate the transition from sporangial to conidial behaviour. For example –

(a) *P. gracile* always gives rise to only zoospores.

(b) *P. debaryanum* produces zoospores, but under dry conditions the sporangium behaves like a conidium i.e., sporangium germinates directly by producing a germ tube.

(c) *P. intermedium* produces both sporangia as well as conidia irrespective of the environmental conditions.

(d) *P. vexans* produces mostly conidia and only occasionally the zoospores are formed.

(e) The conidial condition is finally reached in *P. ultimum,* which always produces conidia.

Hence, different species of *Pythium* clearly illustrate the transformation of a sporangium to a conidium.

Asexual reproduction in *Pythium* is also accomplished by the formation of *Chlamydospores* [Drechsler 1939; 1941]. These thick walled chlamydospores on germination produce a long hypha, terminating in a sporangium which produces zoospores in the usual manner.

Sexual Reproductions

Sexual Reproduction is oogamous. Sex organs are produced at the end of the growing season. Both antheridium and oogonium are produced on the same hypha. *Pythium* is thus *"nomothallic"*.

Oogonium : During the formation of oogonium, the tip of the female branch enlarges into a spherical swelling. Into this swelling a number of nuclei and cytoplasm migrate. It is then cut off from below by a septum. The nuclei already present divide mitotically. As the oogonium matures, the protoplasm becomes differentiated into a central or peripheral denser, rounded portion called *"Ooplasm"*, and an outer or peripheral spongy portion lining the oogonial wall called *"periplasm"*. The ooplasm or oosphere contains a single nucleus, while all other nuclei migrate to the peripheral periplasm and degenerate as in *"P.ultimum"* and *"P.torulosum"* [Trow 1901; Middleton 1927]. In *P.debaryanum,* all nuclei at first migrate to the periplasm and one nucleus later, returns to the *"ooplasm"* and the rest degenerate [Mlyake 1901].

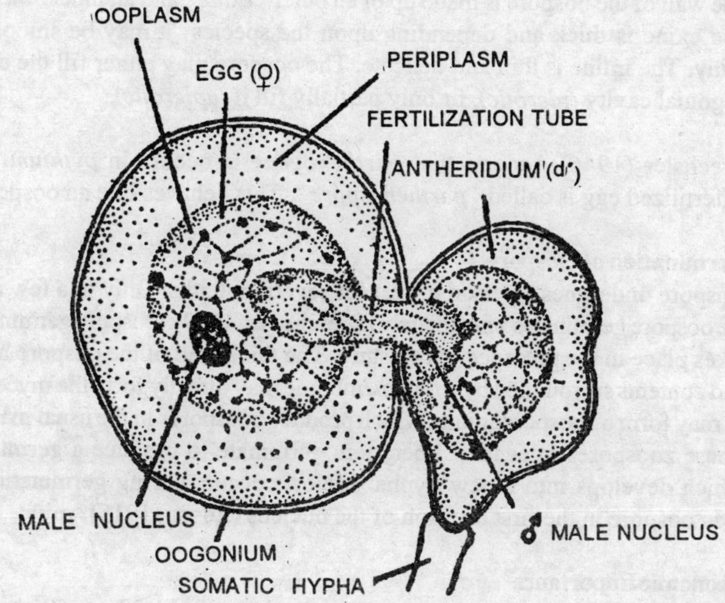

Fig. 11.7 Fungi
Sex organs showing gametangial contact

Antheridium : Antheridium is small and club-shaped; it develops at the tip of the male branch, either on the same hypha which bears the oogonium or on an adjacent hypha. It is cut off from the main hypha by a septum. The antheridium is coenocytic and contains many nuclei. As the antheridium matures, the protoplasmic contents differentiate into a central fertile portion containing a single nucleus called *"male gamete"*, and a peripheral *"periplasm"* containing several nuclei which later degenerate.

Fertilization

The antheridium bends and closely applies itself to the oogonium. One or many antheridia may become attached to a single oogonium. Later, antheridium sends out a small tube, the *fertilization tubei,* or the "*conjugation tube*". This pierces through the oogonial wall and elongates to reach the *"egg"*. The male nucleus enters the oogonium through the fertilization tube and fuses with the egg to form a diploid *"zygote"* or *"oospore"*.

The wall of the oospore is made up of an outer *"exine"* and an inner *"intine"*. The exine is thick and depending upon the species it may be smooth or spiny. The intine is thin and delicate. The oospore may either fill the entire oogonial cavity *(plerotic),* or only partially fill it *(aplerotic).*

Drechsler (1946), reports that *parthenogenesis* occurs in *pythium.* The unfertilized egg is called *"parthenospore"*. This behaves like an oospore.

Germination of Oospore

Oospore undergoes a period of rest for several months. Only in a few cases the oospore germinates immediately [Dreschseler, 1939). Oospore germination takes place in the presence of moisture. The thick wall of the oospore bursts and contents surrounded by a thin membrane may directly form the mycelium or may form a zoosporangium, which produce zoospores in the usual manner. These zoospores after their liberation, germinate to produce a germ tube, which develops into a new hypha. Meiosis occurs during germination of then oospore in the first division of the nucleus (Edson, 1915).

Economic Importance

Several species of *Pythium* have been reported from India. Most of the species of *Pythium* are destructive parasites of economically important plants. They cause the following destructive diseases.

1. Rhizome rot of ginger, or *"soft rot"* of ginger caused by *P.aphanidermatum* and *P.myriotylum.*
2. *Damping off* of tobacco, tomato and chillies, caused by *P.debaryanum.* Venkataraman has reported *P.aphanidermatum* on tobacco seedlings from Mysore.
3. "Soft rot" of papaya has been reported by Subramaniam in 1920. The casual agent is *"P.graminicolum.*

Control of the Disease

Moisture content of the soil and humidity of the air are mainly responsible for the rapid spread of the disease. During these favourable conditions production of zoospores is high, and hence the infection spreads rapidly. Following measures are suggested to prevent the disease:

(1) Lowering the humidity of the atmosphere.

(2) Exposure of the seed beds to air and light.
(3) Prevention of the overcrowding of the seedlings
(4) Sterilisation of the soil by dry heat and steam.
(5) Use of porous and sterile soils.
(6) Treatment of seeds with orango mercuric compqunds.
(7) By the use of chemicals such as Bordeaux mixture and formation etc.

ORDER – MUCORALES

Family – *Mucoraceae*

Mucoraceae is the largest family of this order. The members of the family are found in soil, on decaying plant matter, on dung, or on other fungi etc. Asexual reproduction takes place by sporangiophores or aplanospores. Sporangia posses columella and the sporangial wall is thin. Sexual reproduction takes place by the copulation of gametangia forming zygospores which are naked, or loosely covered by appendages.

Rhizopus and *Mucor* are the two best known genera of this family. *Rhizopus* includes 35 pecies.

PHYCOMYCETES
MUCORALES
MUCORACEAE
RHIZOPUS

Occurrence : *Rhizopus* is commonly called *bread mold* because of its frequent occurrence on stale bread. It lives as a saprophyte. It also occurs as a saprophyte on decaying fruits, vegetables, pickles, jellies and jams etc. *R.stolonifer* grows as a weak parasite on strawberries and sweet potatoes.

Mycelium : Mycelium is cottony white during the vegetative phase. It is evenly distributed in and outside the substratum. The mycelium is profusely branched coenocytic and normally without cross walls. In the older mycelium, plant body shows differentiation of hyphae, consisting of three kinds.

(i) **Stolons :** Stolons are hyphae, which run horizontally over the substrata. They are usually larger and unbranched; their growth is very fast.

(ii) **Rhizoidal hyphae :** These arise from the stolons at definite intervals. They are brown, slender, and branched. These rhizoid like hyphae penetrate deep into the substratum. By their enzymatic action they digest the starchy substratum. Hence, they serve the double purpose of anchoring the fungus to the substratum, and absorb water and nourish the entire plant body.

(iii) **Sporangiophores :** These are the third kind of hyphae developed during the reproductive phase, and are formed just opposite the rhizoidal hyphae. Sporangiophores grow erect and bear sporangia singly at their tip.

Structure of the Thallus

Mycelium is coenocytic and aseptate; septa are formed in connection with the formation of reproductive structure. The wall of the mycelium is made up of *"fungus chitin"*, within the walls of the mycelium is present cytoplasm. Cytoplasm is heterogenous containing numerous nuclei, small vacuoles and droplets of oil and glycogen. Apart from these, hyphae also show brown pigments.

Asexual Reproduction

Asexual reproduction takes place by the formation of small rounded, non-motile spores produced in sporangia which are borne terminally on the negatively geotropic sporangiophores. These reproductive organs develop after a certain period of vegetative activity.

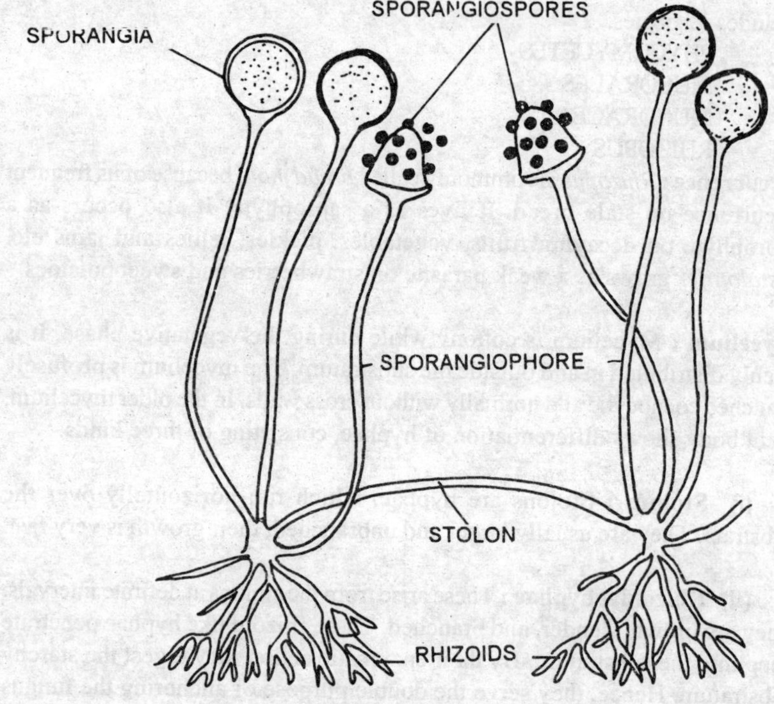

Fig. 11.8 Fungi
Vegetative and asexual structure

During the development of the sporangium, the tip of the sporangiophore enlarges into a spherical sac. Due to the migration of cytoplasm and nuclei from the sporangiophore, this gradually enlarges and at maturity becomes black.

At this stage, the sporangium contains many nuclei and a dense mass of cytoplasm, and reserve food materials. As the sporangium matures, the cytoplasm becomes differentiated into two regions. Most of the cytoplasm and nuclei aggregate at the peripheral region, surrounding a central vacuolar protoplasm containing small number of nuclei. At this stage, between the two regions, small vacuoles appear. These vacuoles mark the region where the spore-bearing protoplasm will be separated from the central region. The vacuoles flatten and finally fuse to form a cleft between the two zones. A wall appears at this region forming a central *"columella"* surrounding an outer *"spore producing region"*.

The outer protoplasmic region undergoes cleavage to form many smaller protopasts, of irregular shape. They later, round off, secrete a new wall and develop into *"spores"*, or *"aplanospores"*. They contain variable number of nuclei.

As the spores mature, columella grows exerting pressure on the sponrangium, with the result the sporangial wall bursts open exposing the spore mass. Walls of the sporangium may become attached at the base of the columella as a rugged collar.

Spores are small and very light. They are mainly carried by the air. In this way, they (spores) become distributed widely and are present everywhere.

Germination of a Spore

Spores of *Rhizopus* are usually globular or traingular and are usually black. These spores are capable of undergoing rest for long periods. When they fall on a suitable substratum like bread, jams, pickles; dung etc., they start germinating. First, spores absorb water and swell. The spore wall ruptures and protoplast emerges in the form of a short tube. This branches and elongates profusely to form a new mycelium.

Rhizopus under unfavourable conditions such as shortage of water etc., reproduces asexually by the formation of *"chlamydospores"*. The chlamydospores are thick walled and can resist the unfavourable conditions effectively. During their information, the hyphae break up the formation of septa. The contents of these cells round off, develop thick walls and form chlamydospores. After the return of favourable conditions, these germinate to form a new mycelium.

Sexual reproduction

Rhizopus may be homothallic *(R. sexualis),* or heterothallic *(R.nigricans).*

Sexual reproduction in *Rhizopus* is remarkable and takes place by the fusion two "coengametangia". This sexual reproduction is called "isogamous". (Gametangial copulation). The product is called *zygospore.* These were first observed by Ehrenberg in 1829. For the formation of zygospores two different strains of thalli are necessary. One of them is called *plus* (+), and the other *minus* (-). These two are morphologically similar. The species having + and – strains are called *heterothallic.* In homothallic species *(R. sexualis),* however, zygospores are formed by the interaction of hyphae arising from the same spore. This phenomenon of *heterothallism* was first noticed by Blackslee (1904).

During the process of sexual reproduction, two compatible hyphae (+ and –) are attracted towards each other. Each of these hyphae produces a short copulating branch called *"Progametangium".* Progametangia are produced by each hypha at the point of contact. These progametangia gradually enlarge and a transverse septum arises at the tip forming a small terminal cell called the *"gametangium".* The remainder of the progametangium is now called *"suspensor".* Gametangium contains dense and multinucleate protoplasm. The progametangia are usually similar when they are called *"isogamous",* and are called *"anisogamous"* when they are unequal. At maturity, walls between the two gametangia dissolve and the cytoplasm and nuclei mingle. Nuclei of opposite strains fuse forming a large number of diploid nuclei. This combined protoplast is now called *"zygospore".* The suspensors gradually wither away setting free the zygospores.

Zygospore gradually increases in size and secretes a thick wall which is usually two layered. The outer wall called *"exine"* or *"exospore"* is thick, dark and warty. The inner layer is thin and is called *"intine"* or *"endospore".* Before germination zygospores undergo a period of rest.

Germination of Zygospore

Germination of Zygospore takes place under favourable conditions. At the time of germination, zygospore absorbs moisture and swells up. Due to this swelling, exposure ruputers and endospore with the protoplast protrudes in the form of a branched tube. This is called the *"promycelium".* Promycelium grows to a limited extent. This at its tip bears a typical *"zygosporangium",* or the "germ sporangium". During germination, the diploid nuclei undergo reduction division and there is the segregation of + and – strains. Later, zygosporangium bears haploid spores. When these spores fall on a suitable substratum they germinate to form a new mycelium of *Rhizopus*.

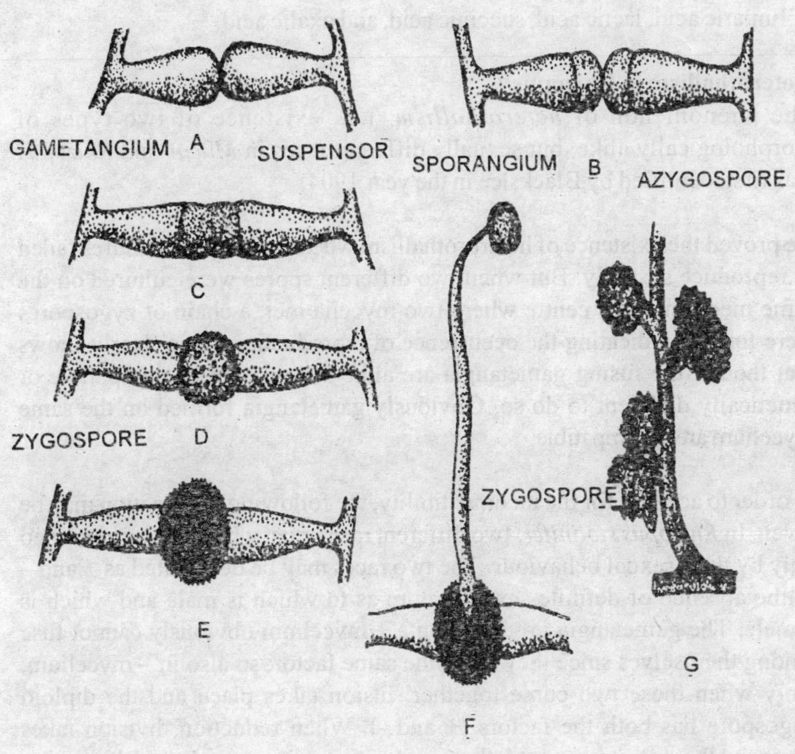

Fig. 11.9 Fungi
Sexual reproduction. Various stages in the development of the Gamentangia and Zygospore

Parthenospores : Sometimes the nuclei fail to fuse and develop without conjugation into thick walled spores. These spores are called *"parthenospores"* (azygospores). These behave like ordinary zygospores.

Economic importance

Though the black molds are mostly saprophytic, some of them are pathogenic.

Rhizopus arrhizus causes great loss to young apples. *R.nigricans* (syn. *R.stolonifer),* causes a serious transit disease of strawberries called *"leak"*, and a *"soft rot"* of sweet potatoes in storage. *Rhizopus oryzae* together with *Mucor praini* and *Mucor javanicus* saccharify starch and can be used in fermentation.

Various species of R*hizopus* and *Mucor* are capable of forming large quantities of fumaric acid, lactic acid, succinic acid, and oxalic acid.

Heterothallism in Rhizopus

The phenomenon of *heterothallism,* i.e., existence of two types of morphologically alike, but sexually different races in *Mucor* and *Rhizopus* was first identified by Blackslee in the year 1904.

He proved the existence of heterorothallism when single spore cultures failed to reproduce sexually. But when two different spores were cultured on the same medium at the centre where two mycelia met, a chain of zygospores were formed indicating the occurrence of reproduction. This clearly shows that though the fusing gametangia are alike they should be compatible or genetically different to do so. Obviously gametangia formed on the same mycelium are incompatible.

In order to account for the incompatibility, the following explanation may be given. In *Rhizopus stolinifer,* two different races exist which can be identified only by their sexual behaviour. The two races may be designated as + and – in the absence of definite sexual nature as to which is male and which is female. The gametangia produced on a + mycelium obviously cannot fuse among themselves since they carry the same factor; so also in – mycelium. Only when these two come together, fusion takes place and the diploid zygospore has both the factors (+ and -). When reduction division takes place in the zygospore and the germ sporangium produces the spore, segregation of factors takes place and 50 percent of the spores are of + type and the other 50 – type. These develop into corresponding mycelia. At the spore stage, however, it is not possible to distinguish the two. Only by observing their sexual behaviour can one identify them. This phenomenon is known as *heterothallism.*

ASCOMYCETES
ENDOMYCETALES
SACCHAROMYCETACEAE
SACCHAROMYCES (YEASTS)

The family includes about 10 genera and 100 species. The members are commonly called *"Yeasts"*. Yeasts are unicellular organisms. They have the capacity to ferment sugar into ehtyl alcohol and carbon dioxide. Yeasts reproduce asexually by means of "fission" or "budding", or by means of both methods. Zygote functions directly as an ascus. Hence, yeasts do not produce definite *"ascocarp"*.

Occurrence : Yeasts are widely distributed in nature. They are saprophytes, and are commonly found on sugar substrata like nectar of flowers, on the surface of fruits like grapes, apples etc. Yeasts also occur in milk and other foods, subject to fermentation. They are also of common occurrence on the excreta of animals. Some yeasts are pathogenic on plants causing leaf curl disease of various species.

Culture of Yeasts : Yeasts can be easily cultured in the laboratory by placing small pieces of yeast cake in sugar solution. The yeasts cells multiply and a profuse growth of yeasts may be obtained within one or two days.

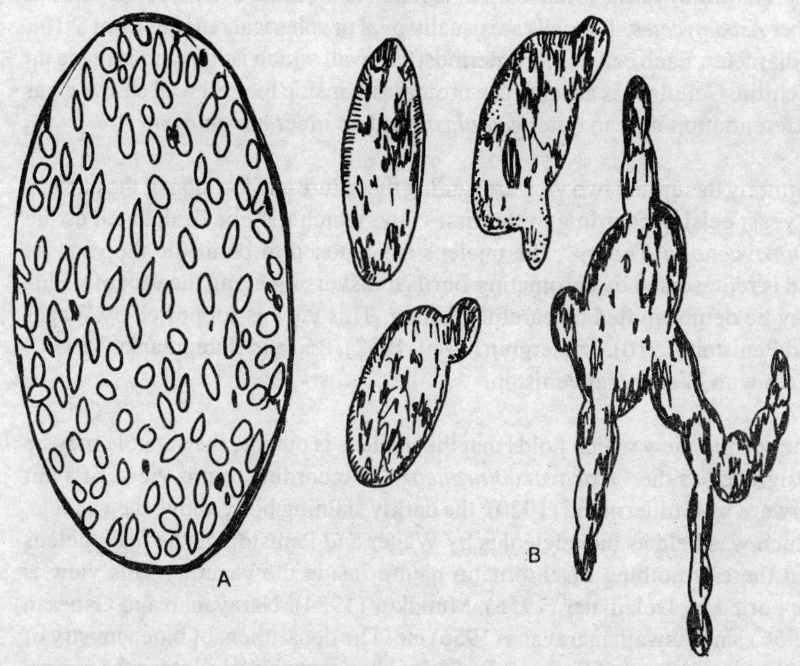

Fig. 11.10 Fungi

A. Yeast cells under microscopic (Medium power view), **B.** Enlarged yeast cells.

Fermentation

Yeasts are very well known for the process of fermentation. They have the capacity of fermenting carbohydrates. Hence, they are called *saccharomycetes (Saccharon* = Sugar, *mykes* – fungi).

Fermentation of sugar substrate is by the action of an enzyme called *"Zymose"*. In addition to the enzyme, certain nitrogenous salts and phosphate are essential. During the process of fermentation, sugars are broken down by the enzyme zymase into carbon dioxide and ethyl alcohol. During this process small traces of acetic acid, succine acid, and glycerine are also formed. The process of fermentation can be represented as follows :

$$\underset{\text{(Sugar)}}{C_6H_{12}O_6} \rightarrow \underset{\text{(Ethyl alcohol)}}{2C_2H_5OH} + \underset{\text{(Carbon-dioxide)}}{2CO_2}$$

The process of fermentation takes place when there is insufficient supply of oxygen.

Cell structure

The Thallus in yeasts is made up of isolated cells, and not a mycelium like in other *Ascomycetes*. The cells are usually oval or spherical ranging from 5-10μ in diameter. Each cell has an outermost cell wall which isthin, and is made up of chitin. Cellulose is absent. The protoplasm inside the cell wall shows clear differentiation into an outer *ectoplasm* and an inner *endoplasm*.

Formerly there were two views regarding the nature and location of the nucleus in yeast celsl. According to the first view, which may be designated as the *"Intravacuolar Theory"*, the nucleus is supposed to be inside the vacuole and is represented by chromating fibrils transversing along the vacuole. This may be demonstrated by careful staining. This view is supported by Wager and Peniston (1910). Lindergren (1949, 1952), Rao and Subramanian (1953), agree with Wager and Peniston.

The second view which holds that the nucleus is outside the vacuole may be designated as the *"Extravacuolar theory"*. According to this view, first put forward by Guillermond (1920), the darkly staining body above the vacuole, which was held as the nucleolus by Wager and Peniston, is the real nucleus and there is nothing of chromatin nature inside the vacuole. This view is supported by Delamater (1955), Mundkur (1954), Narayanan and Ganesan (1956), and Aswathanarayana (1956) etc. The department of biochemistry of the Indian Institute of Science, Bangalore has contributed a lot for the correct understanding of the yeast cell structure and the findings greatly support the extravacuolar theory.

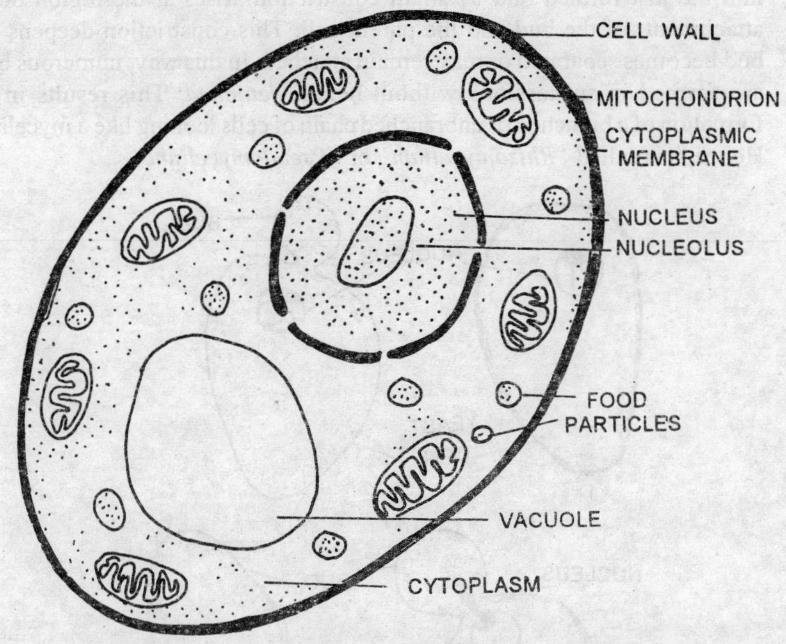

Fig. 11.11 Fungi
Detailed cell structure (Electron microscopic view)

Cytoplasm contains reserve food materials like glycogen and oil. It also contains mitochondria, starch grains etc.

REPRODUCTION

Reproduction takes place by means : (1) *Vegetative* (2) *Asexual and* (3) *Sexual methods.*

Vegetative Reproduction

Vegetative reproduction is brought about by means of (a) *budding* and (b) *fission.* Budding is the common method of vegetative reproduction in *Saccharomyces.*

Budding

Budding takes place when there is abundance of food material. In some of the yeasts, this is the only method of reproduction. During the process of budding, a small bulging arises from one side of the cell. This gradually increases in size. This is called a *bud.* Meanwhile, the nucleus of the mother cell divides into two daughter nuclei, one of these daughter nuclei migrates into the just formed bud. A small constriction arises at the region of the attachment of the bud and the parent cell. This constriction deepens and bud becomes separated or may remain attached. In this way, numerous buds are formed in successions without being, *detached.* This results in the formation of a branched or unbranched chain of cells looking like a mycelium. Hence, it is called *"Rhizomycelium"* or *"Pseudomycelium"*.

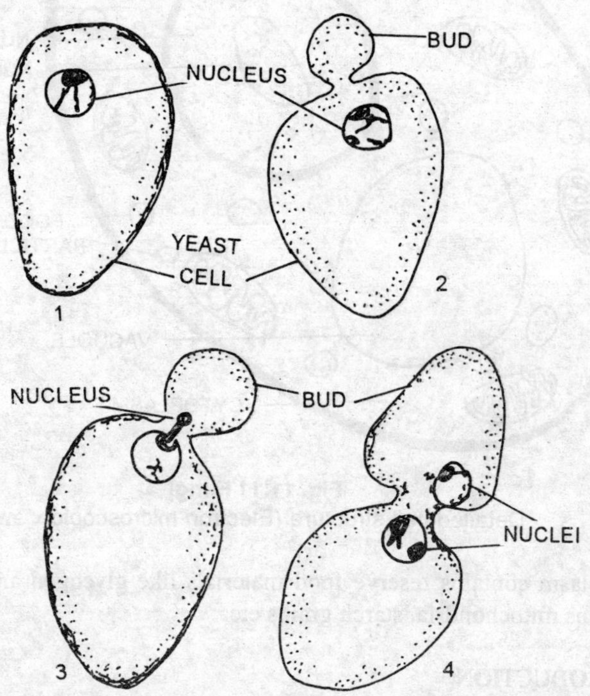

Fig. 11.12 Fungi
Asexual reproduction by budding

Fission

Fission is a common occurrence in *Schizosaccharomyces octosporus.* During fission, the mother cell elongates, nucles divides into two daughter nuclei, and this is followed by the formaton of a transverse septum which deepens and two daughter cells are ultimately formed. These are of equal size, each containing the normal components of the cell. The cells separate to lead an independent life.

(2) Asexual Reproduction

This takes place during unfavourable conditions i.e., when there is a shortage or complete lack of food supply. During these conditions protoplast of the cell recedes from the cell wall and cytoplasm collects around each nucleus and develops into thick walled spores, called *"endospores"*. Later, these are liberated. Under favourable conditions these give rise to chains of buds, which separate and become independent. If conditions are not favourable, they undergo a period of rest.

Sexual Reproduction

Sexual reproduction in yeasts is of a very simple type. It occurs during unfavourable conditions. Sexual fusion takes place between two somatic cells, or between two ascospores which act as *"gametangia"*. Two gametangia come together and fuse to form a diploid zygote. Zygote directly becomes an ascus and produces ascospores after reduction division. The number and shape of ascospores, varies. Usually 4 or 8 ascospores are liberated by the rupture of the ascus. After their liberation, they divide by means of fission or budding to form somatic cells.

Life cycles in Yeasts

Guillermond (1940), has recognised three different types of life-cycles in Yeasts, namely–

(1) Haplobiontic Eg. *Schizosaccharomyces octosporus.*

(2) Diplobiontic Eg. *Saccharomycodes ludwigi.*

(3) Haplo diplobiontic Eg. *Saccharomyces cerevisiae.*

(1) Haplobiontic Life-Cycle

This type of life-cycle is exhibited by *Schizosaccharomyces octosporus.* Haploid stage here, is of greater duration, while diploid stage is represented only for a short period hence the life-cycle is called *"Haplobiontic"*. In this type, zygote which is diploid stage, undergoes meiosis to form 8 ascospores. Hence, here zygote directly acts as an ascus bearing 8 ascospores. After their liberation ascospores behave as vegetative cells, increasing their number by means of fission. During copulation, any of the two cells come together, become opposed end to end. They produce small outgrowths. These protruberances fuse forming a *"conjugatin tube"*. Nucleus from each cell passes into this tube and fusion takes place to form a diploid zygote. This zygote behaves as explained above, and the cycle continues.

Hence, in this type of life-cycle haploid stage is of longer duration, while diploid stage is of short duration, being represented only by the zygote.

Diplobionic life-cycle seen in *Saccharomycodes ludwigii*

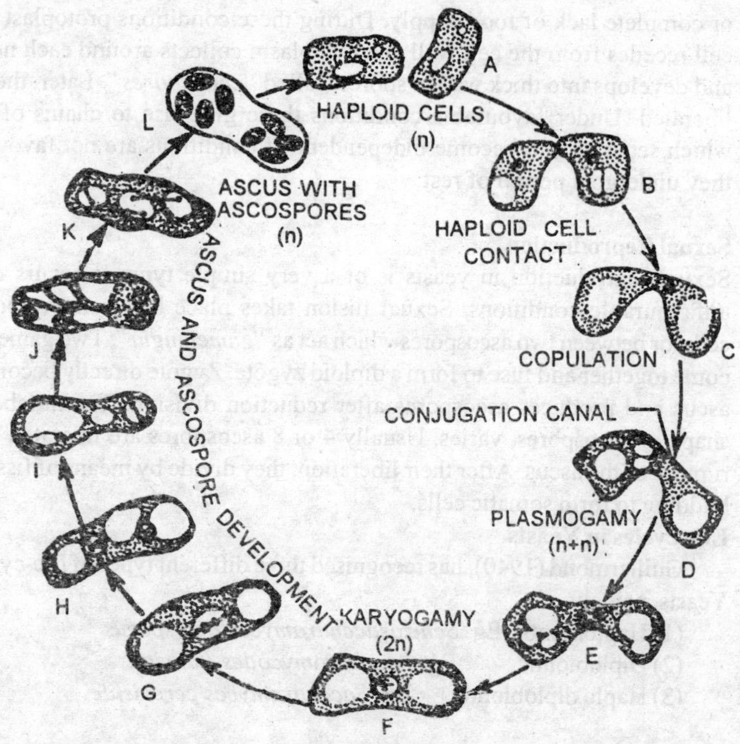

Fig. 11.13 Fungi
Haplobiontic life cycle seen in Schizosaccharmyces octosporus

(2) Diplobiontic Life cycle.

This type of life-cycle is exhibited n *"Saccharomycodes ludwigii"*. In this case, diploid stage is of longer duration, hence the life cycle is called *"Diplobiontic"*.

Here, only 4 ascospores are produced after meiosis in the zygote (ascus). These four ascospores are not released, instead, they fuse among themselves to form 2 diploid zygotes, which lie within the ascus. Each zygote, produces a short germ tube, this pushes the ascus wall and germinates into *"Sprout mycelium"*. This, buds off small, diploid yeast to produce four ascospores. Hence, in this life cycle haploid stage is very short and is represented only by the 4 ascospores.

(3) Haplo-Diplobiontic life-cycle

This type of life-cycle is seen in *Saccharomyces cerevisiae.* In this case, both haploid and diploid stages are represented equally well in the life-cycle hence the life-cycle is called *"Haplo diplobiontic"*. During fusion two haploid cells come nearer and fuse to form a diploid cell. This multiplies by budding, and produces large numbers of diploid cells. Later on, these behave as asci, producing 4 ascospores after reduction division. These ascospores are liberated to the outside by the rupture of the ascus wall. These again multiply by budding to produce large numbers of haploid vegetative cells, which again fuse and repeat the life-cycle.

Hence, in this type of life-cycle haploid and diploid stages are of equal duration.

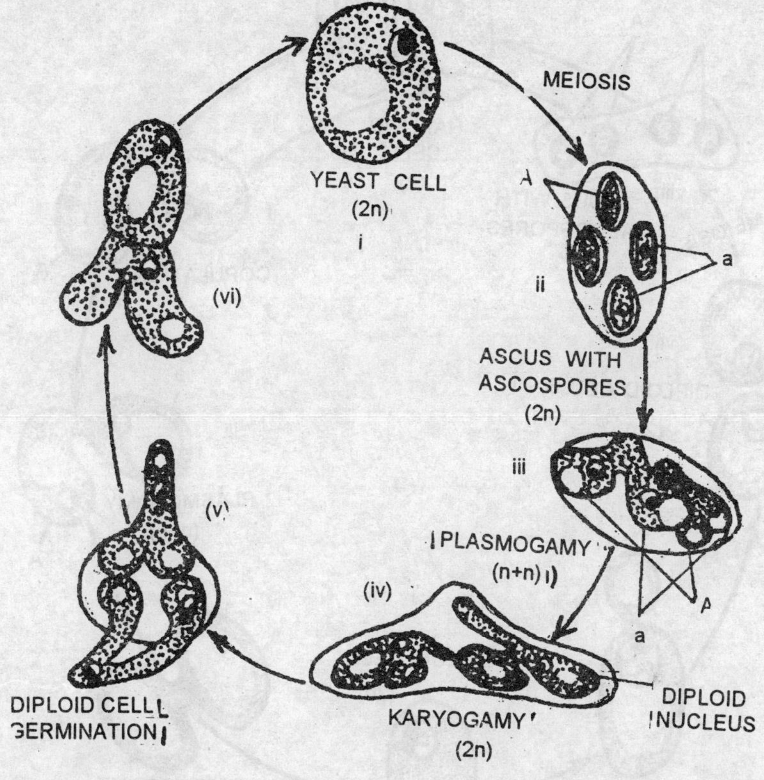

Fig. 11.14 Fungi
Diplobionic life cycle seen in Saccharomycodes hudwiggi

Economic Importance

The main use of yeasts is in *baking* and *brewing* industries. As already mentioned, the importance of yeasts lies in their capacity to ferment sugary substrate into ethyl alcohol and carbon dioxide. These two products are used by brewers and bakers respectively. In the brewing industry, alcohol is the important industrial product and carbon dioxide is a waste; whereas in the baking industry, carbon dioxide is the important product and alcohol is a waste. Different types of yeasts give different flavours to the alcohol. Carbondioxide used in the baking industry helps in raising of the flour and hence the bread becomes spongy.

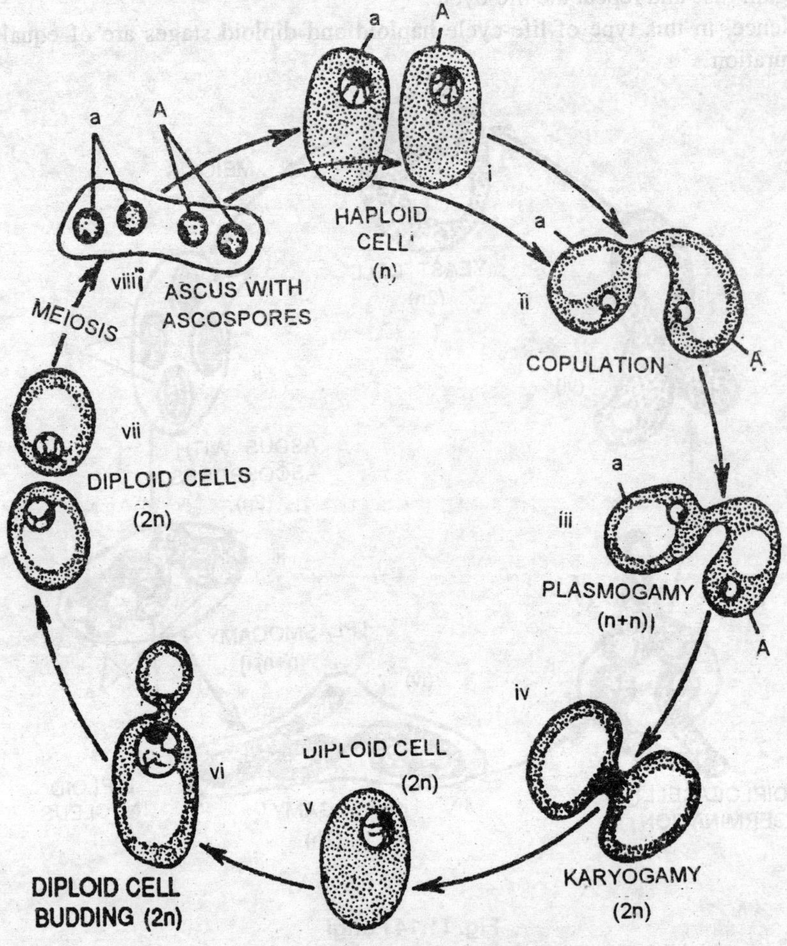

Fig. 11.15 Fungi
Haplodiplobiontic life cycle seen in Saccharomyces cerevisiae

Due to their immense use in baking and brewing industries, yeasts are used in the preparation of *Yeast cakes* on large scale. During the preparation of yeast cakes, large number of yeasts are pressed together with an inert matter such as starch. Later on, they are cut into cubes, dried and sold commercially as "Yeast cakes". Because of their high vitamin (B and C) protein and fat content, yeasts are also used as food.

Apart from these beneficial uses yeasts cause certain skin and intestinal diseases and are repsonsible for the spoilage of food substances.

ORDER-ASPERGILALES

Family-Aspergillaceae

The Family- Aspergiallaceae is characterised by the formation of *Cleistothecia* with definite *peridia*. Peridium is made up of compactly arranged pseudoparenchyma. Conidial formation is very characterisitic of this family. Conidia when mature are echinulate. Immature ones are smooth walled. The family includes many genera, among which *Pencillium* and *Aspergillus* are very important.

ASCOMYCETES
ASPERGILLALES
ASPERGILLACEAE
PENICILLIUM

Occurrence : *Penicillium* is commonly called *green* or *blue mold,* and is a saprophytic fungus, which grows on rotten fruits, rotten vegetables, meat etc. Many of the species are of great economic importance. *P.italicum* and *P.digitatum,* infect fruits belonging to family Rutaceae, especially citrus fruits. Infected fruits show greenish or slightly bluish coating which is due to the presence of conidia.

P.expansum, causes great damage to apples, pears and grapes in storage. Apart from these, many species are also beneficial to human beings. These will be discussed later under the economic importance of *Penicillium.*

Mycelium : The mycelium is slender, composed of profusely branched septate hyphae. The septa have a central opening through which cytoplasm move from a cell. Some of the hyphae enter deep into substratum and help in absorption. The wall is thin and cells are multinucleate. In *P.vermiculatum* the cells are uninucleate. In some species, mycelium may form sclerotia.

Reproduction

Penicillium reproduces vegetatively, asexually and sexually.

Vegetative reproduction is brought about by the fragmentation of hyphae into a number of fragments. Each fragment develops into a new mycelium.

Asexual Reproduction

Asexual reproduction is the most common method of reproduction in case of *Penicillium.* It takes place by the formation of *conidia* on branched *conidiophores.* These conidiophores arise from vegetative mycelium and not from the specialized foot cell as in Aspergillus. These are erect and branched once or twice. Conidiophores are unbranched in *P. thomii.* Each conidiophore at its tip bears a single whorl of bottle-shaped structures which are called *"Sterigmata"* or "Phialides". The first formed conidiophores called "primary branches" may again branch to form secondary or even tertiary branches. The ultimate branches which bear sterigmata are called *"metulae".* They are of systematic value. Each sterigma at its tip cuts off conidia in basipetalous succession.

During the formation of conidium, the tip of the conidiophore enlarges into a spherical structure. Meanwhile, the nucleus divides into two. One of them migrates into the just formed swelling. This is later cut off by the formation of a transverse septum, develops a wall of its own and becomes a conidium. This process continues and several conidia may be seen attached at the tip of the conidiophore one below the other. The oldest conidium is towards the tip and youngest towards the base. Hence, the development is called basipetalous succession. The entire sterigmata, and chains of conidia resemble a small brush or broom. Hence, the structure is technically called *Penicillus,* which means a brush in Latin. Such brush like conidiophores are characteristic of *Penicillium.*

Conidia when liberated are usually spherical with a bluish or greenish colour depending upon the species. These are usually uninucleate, but may become multinucleate in some species.

Upon failing on a suitable substratum, conidia germinate into a fresh mycelium by producing a short germ tube.

Sexual Reproduction

Sexual reproduction is not commonly found in *Penicillium.* It is observed only in about 20 species. Sexual reproduction in *P.glaucum,* was first studied by Brefeld (1974). Dangeard (1907), noticed sexual reproduction in *P.vermiculatum.* In this species, sexual reproduction is oogamous. Male sex-organs are called *"antheridia",* and female sex organs are known as *"ascogonia".*

Ascogonium

Ascogonium, is elongated flask, shaped and uninucleate at the beginning.

As it matures, the nucleus divides to form about 64 nuclei. Ascogonium at this stage is without septa.

Antheridium

A slender hypha originates generally from a different but adjacent hypha than that which bears the *"ascogonium"*. This slender branch elongates and twines round the ascogonium several times. This anthridial branch at its tip cuts off *antheridium* proper by a transverse septum. The *antheridum* at this stage, is club shaped and uninucleate.

The antheridium comes in contact with the ascogonium. At this region of contact, walls dissolve and theprotoplasts of the antheridium and ascogonium mingle with each other. According to Dangeard, there is no fusion of the two protoplasts, hence there is no *gametic union,* instead, the antheridial protoplast degenerates. Further, according to him the nuclei in the ascogonium arrange in pairs and ascogonium becomes divided by means of transverse septa. Each cell at this stage contains two nuclei, derived from the ascogonium. This phase is called *Dikaryotic phase,* and each pair of nuclei is called a *"Dikaryon"*.

In *P.glaucum* sexual reproduction has been studied by Brefeld (1874). It takes place by the formation of two copulating branches; one of which is the *"antheridium,* and the other forms the *"ascogonium"*. *Plasmogamy* takes place by *gametangial contact.* Dikaryotic phase is formed by the septation of the ascogonium into binucleate cells, which later develop into ascogenous hyphae and finally into ascocarp.

Development of Ascocarp

Ascocarp in case of *Penicillium* is a *clestothecium.* During its formation, some of the binucleate cells become *ascogenous initials.* Each ascogenous initial develops into a branched *ascogenous hypha* composed of two nucleate cells. During the formation of ascus, the binucleate cell, usually situated at the terminal region becomes the *"ascus mother cell"*. In this, karyogamy occurs forming a diploid nucleus. The young ascus enlarges, meanwhile, the diploid nucleus undergoes 3 divisions to form 8 haploid ascospores. The first two divisions are meiotic, while the third division is mitotic. Meanwhile, large number of sterile hyphae grow around the ascus and completely encircle it to form the *"ascocarp"*.

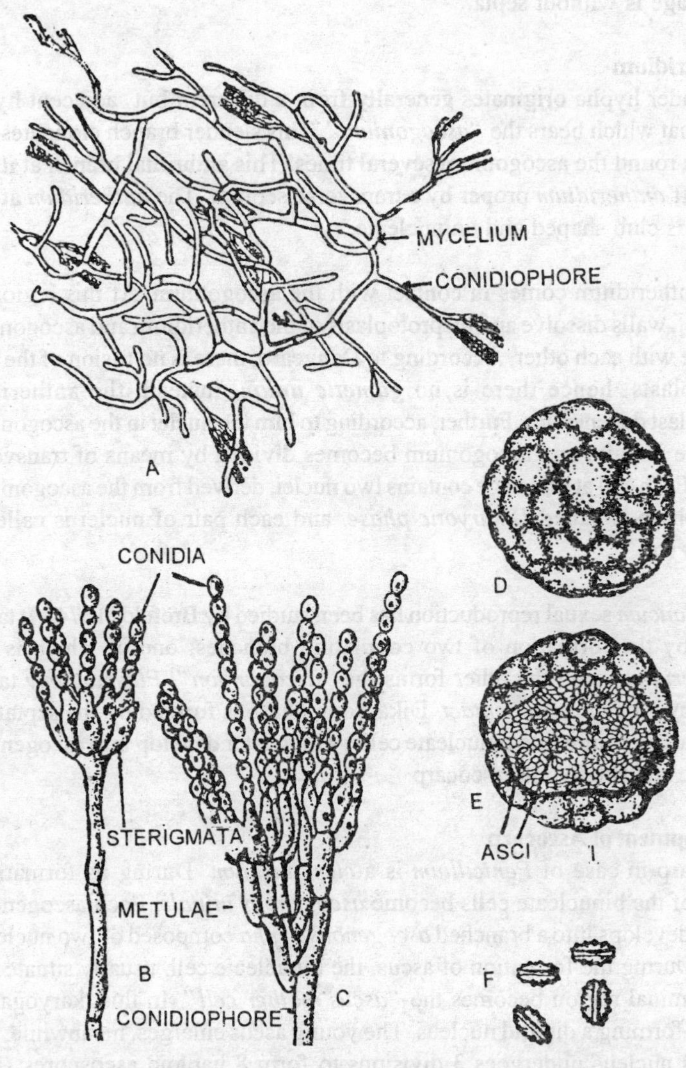

Fig. 11.16 Fungi

A. Somatic structure, **B-C.** Enlarged condiophores, **D.** Cleishothecium, **E.** Cleistothecium showing Asci, **F.** Ascospores

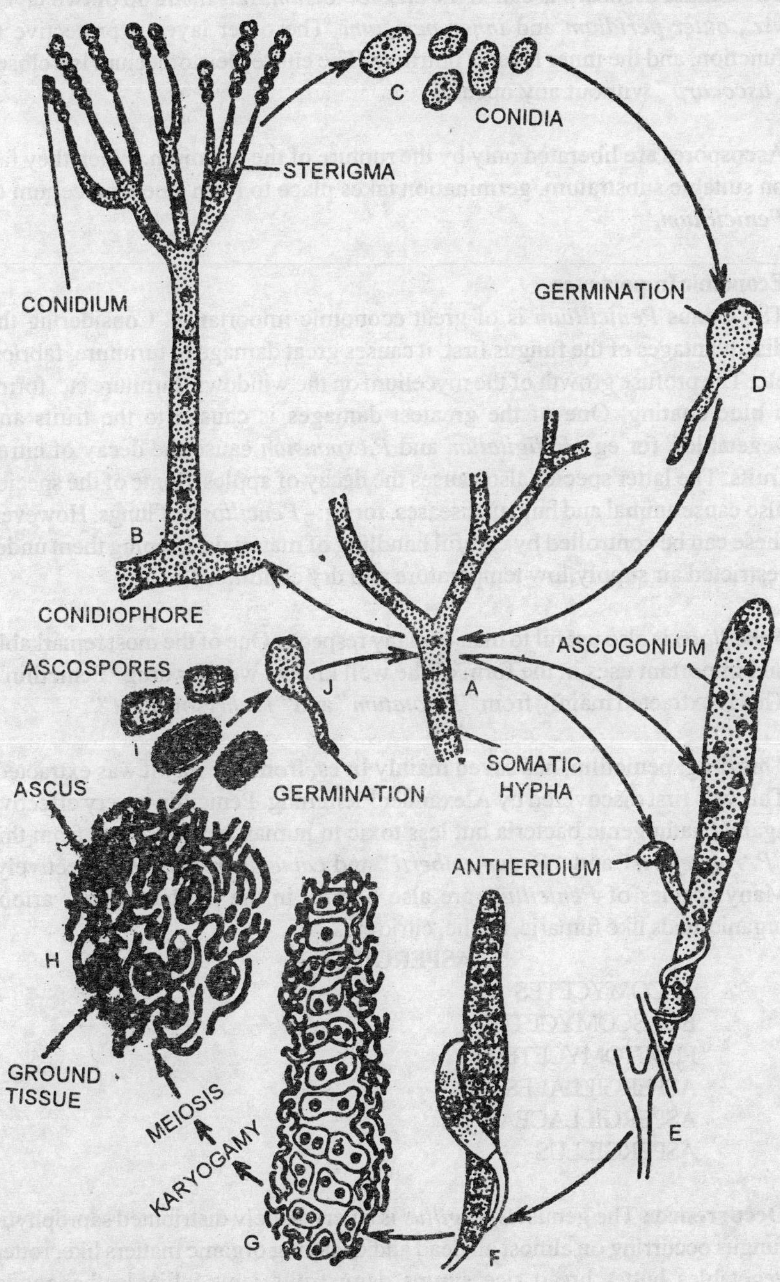

Fig. 11.17 Fungi
Penicillium. Pictorial representation of the life cycle

The mature ascocarp is called the *cleistothecium*. It is made up of two layers viz., *outer peridium* and *inner peridium*. The outer layer is protective in function, and the inner layer is nutritive. The entire cleistothecium is a closed *"ascocarp"* without any opening.

Ascospores are liberated only by the rupture of the peridium. When they fall on suitable substratum, germination takes place to form a new mycelium of *Penicillium*.

Economic Importance

The Genus *Penicillium* is of great economic importance. Considering the disadvantages of the fungus first, it causes great damage to furniture, fabrics, etc. The profuse growth of the mycelium on the windows, furniture etc, forms a blue coating. One of the greatest damages is caused to the fruits and vegetables, for eg:–*P.digitatum* and *P.expansum* cause the decay of citrus fruits. The latter species also causes the decay of apples. Some of the species also cause animal and human diseases, for eg:– *Pencillosis* of lungs. However, these can be controlled by careful handling of materials, keeping them under restricted air supply,low temperature and dry conditions.

Pencillium is also useful to man in many respects. One of the most remarkable and important uses in the form of the well known wonder drug "Penicillin". This is extracted mainly from *"P.notatum"* and *"P.chrysogenum"*.

This drug, penicillin, has saved mainly lives, from the day it was extracted. This was first discovered by Alexander Flemming. Penicillin is very effective against pathogenic bacteria but less toxic to human beings. Apart from this *"P. roqueforti"* and *"P. camemberti"* and *camembert* cheese respectively. Many species of *Penicillum* are also helpful in the production of various organic acids like fumaric, oxalic, citric etc.

ASPERGILLUS

ASCOMYCETES
EUASCOMYCETES
PLECTOMYCETES
ASPERGILLALES
ASPERGILLACEAE
ASPERGILLUS

Occurrence : The genus *Aspergillus* is a very widely distributed saprophytic fungus occurring on almost all dead and decaying organic matters like, rotten vegetables, butter, bread, rice, syrups, damp fruits, jams, jellies leather goods, rotten wood etc. According to Raper and Fennel (1965), about 200 species have been recognised in the genus of which 30 species have been recorded from India. Sarbhoy (1985), identifies about 132 species in *Aspergillus*. Many species occur in soil also. Some species (*A. flavus* and *A. fumigatus*), cause diseases of human beings (Aspergillosis), and animals. *A..flavus* causes the disease of the human ear called *Otomycosis*. *A. niger* is a very common contaminant of food stuff.

Laboratory culture : A piece of moist bread kept in a petridish and covered with bell jar is the easiest method of culturing *Aspergillus* in the laboratory. Greenish yellow or black patches of cottony mycelia will emerge in a few days. *A nigeri* group forms black molds while *A. glaucus* group forms greenish blue molds.

Somatic structure : The plant body consists of numerous interwoven mass of thread like structures (mycelia), which are profusely branched. The branches (hyphae) of the mycelium grow on the substratum as well as grow into it, absorbing the nutrition. The hyphae are multi celluar with clear septa demarcating the cells. Each cell of a hypha is surrounding by a thin wall enclosing granular cytoplasm. Man nuclei are distributed in the cytoplasm. Other cytoplasmic inclusions found are mitochondria, endoplasmic reticulum and ribosomes, besides oil globules as reserve food material. Each septum has a central pore to facilitate the migration of nucleri and cytoplasm from one cell toanother.

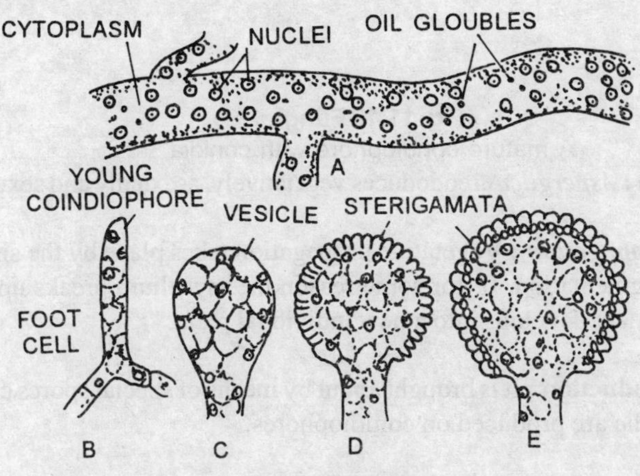

Fig. 11.18 Fungi
Aspergillus. **A**. A portion of hypha; **B-E**. Development of vesicle

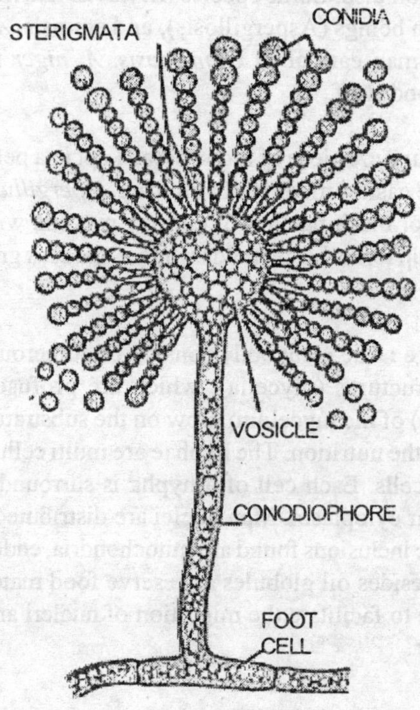

Fig. 11.19 Fungi
A mature condiophore with conidia

Reproduction : *Aspergileus* repdoduces vegetatively, asexually and sexually.

Vegetative Propogation : Vegetative propagation takes place by the simple process of fragmentation. If, for some reason the mycelium breaks up into bits, each fagment develops into a new individual plant.

Asexual reproduction : It is brought about by means of special spores called conidia. Conidia are produced on conidiophores.

Conidiophores are erectunbranched hyphae that arise from the horizontal hyphae. The cell that gives rise to a conidiophore is called the *foot cell.* The condiophores are long aseptate and are free, and terminate in a larger globular structure called vesicle. The vesicle is not separated from the conidiophore by any septa. The conidiophore as well as the vesicle have a large amount of cytoplasm and multinucleate with minute vacuoles. The entire surface of the globular vesicle produces a number of structures called *phialides* or *sterigmata.* Each sterigma is bottle shaped or finger shaped having an accentuated base and tapering apex. There are a number of nuclei in each sterigma, and it is not separated from the vesicle by a septum. Usually, there will be only one layer or series of sterigma arising from the vesicle. In some cases, however the first layer of sterigmata cut off one more series of sterigmata above them; in which case the sterigmata attached directly to the vesicle are called *primary sterigmata* and the ones above them are called *secondary sterigmata.*

Development details of sterigmata : Hanlin (1976), has studied the process in detail of *A. davatus.* At the time of formation of sterigmata, many thin areas appear in the periphery of the vesicle due to the dissolution of the wall material. At these areas, the cytoplasm projects out and forms the sterigmata. Each sterigma also gets a nucleus from the vesicle.

Development of conidia : At the tip of each sterigma, many globular unicellular, uninucleate or multinucleate spores called conidia are formed. Each conidium is back, brown or yellow coloured and has a rough wall. Conidia develop in basi petalous succession (ie., youngest will be at the base and oldest will be at the top).

Development of conidia in great detail has been studied by Subramanian (1971), in *A.niger.* As the conidia develop from phialides (sterigmata), they are also called *Phialoconiada.* In *A.niger,* the sterigma is uninucleate, but in some species it could be multinucleate also.

At the time of formation of a conidium, the tip of the sterigma undergoes a constriction and forms a knob like projection. Meanwhile, the nucleus in the sterigma divides and one nucleus enters the knob like projection along with some cytoplasm. The knob like projection is called the conidium. Meanwhile, the nucleus in the sterigma divides again producing two daughter nuclei. At the same time the sterigma forms one more knob like projection below the first formed conidium, into which enter some cytoplasm and a single nucleus. This knob like projection forms the second conidium. In this way, a number of conidia in chains are formed. As the first formed conidium is gradually pushed up it will be top most and the youngest conidium will be nearest to the sterigma. This type of development is known as basipeitalous.

In the young condition, conidia maintain cytoplasmic connection between them through a constriction called *Isthmus* or *connective.* But as the conidia mature, a cross wall develops between the conidia. The conidia are dispersed by wind and on the availability of a suitable substratum they develop into a new mycelium.

Electron microscopic studies conducted on the development of sterigmata and conidia in *A. giganteus* (Trinci etal, 1968), and *A. clavatus* (Hanlin, 1976), have revealed the process of cross wall formation between the conidia.

Sexual Reproduction

Sexuality in *Aspergillus* is an interesting aspect of study as several species seem to have given up sexual reproduction and reproduce only asexually. In still others, either the antheridium or both antheridium and ascogonium are improperly developed. Sexual reproduction, however, has been reported in a few species.

The sexual stages (perfect stages), which were once thought to be a different genera of Ascomycetes before their relationship with *Aspergillus* was established, have now been included under different species of *Aspergillus* (For example *Eurotium repens* = *Aspergillus glaucans, Sartorya fumigatus* = *A fumigatus).*

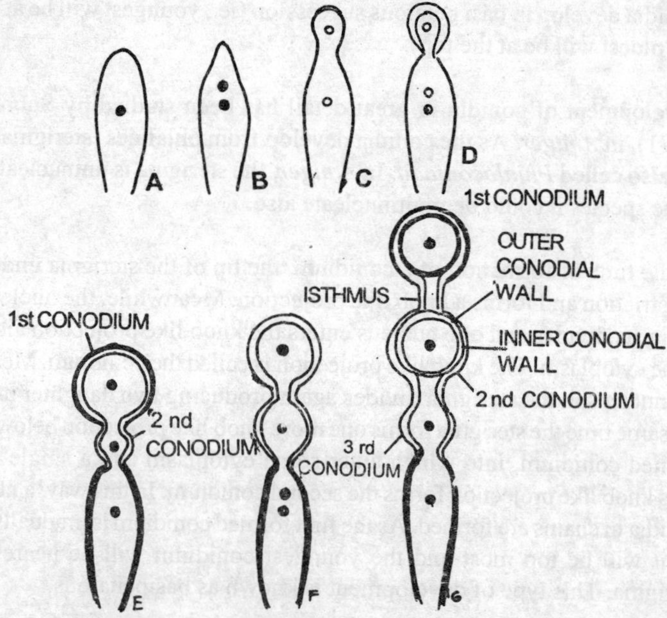

Fig. 11.20 Fungi
A-G. Aspergillus niger, showing development of conidia from a phalide

Sexual reproduction when present is of the oogamous type. The sex organs are antheridia (male), and ascogonia (female). Most of the species are homothallic (both the sex organs develop on the same mycelium). Only a few species like *A. fischri,* and *A.heterothallicus* are heterothallic (Chung and Sang, 1974).

Antheridium : The antheridium first develops as a special branch from the hypha which produces the female sex organ or from another hypha of the same mycelium. The male branch is called the *pollinodium.* This grows nearer to the female sex organ and there, its tip cuts off a unicellular antheridium. The lower portion is now called the stalk. The antheridium is multinucleate.

Ascogonium : This arises as a separate tightly coiled hyphal branch from the mycelium. The branch which subsequently gives rise to the ascogonium is called the *archicarp.* The archicarp is distinguished into three parts – (i) a basal multicellular stalk, (ii) a middle unicellular asogonium, and (iii) a terminal unicellular trichogyne. The trichogyne is the elongated neck like structure meant to receive the male nuclei. All cells of the archicarp are multinucleate.

Fertilization : The tip of the trichogyne comes into contact with the antheridium. The walls between the two dissolve and the antheridial contents flow into the ascogonium. The two cytoplasms (male and female), fuse. This is known as plasmogamy. Subsequent development stages are not clearly worked out. The male nuclei pair with the nuclei of the ascogonium. This is called *dikaryotization.* At this stage, the ascogonium gets divided into a number of cells with each cell having a dikaryon (one male nucleus & one female nucleus). From each dikaryotic cell an outgrowth called ascogenous hypha develops. Each ascogenous hypha becomes multinucleate due to the division of its two nuclei. Ascogenous hypha become septate with the terminal cell becoming uninucleate. The penultimate cell (i.e. cell below the tip cell), which has two compatible nuclei functions as the ascus mother cell. Miller (1984), who has studied development of ascocarp in *Aspergillus* confirms that out of the two nuclei of the ascus mother cell, one is male and the other is female. According to some mycologists, the terminal cell itself is binucleate and it functions as the ascus mother cell.

In any case, the two nuclei fuse in the ascus mother cell resultng in a diploid nucleus. Meanwhile, the ascus mother cell forms a protrusion into which is transferred the diploid nucleus. The diploid nucleus undergoes reduction division followed by mitosis to form eight haploid nuclei. Each nucleus surrounds itself with a little cytoplasm and becomes an ascospore.

Development of ascocarp : At the time when ascogenous hyphae and asci are developing, many sterile hyphae grow up from the base of ascocarp and enclose the asci all around. The sterile hyphae get interwoven with each other and form a pseudoparenchymatous wall called peridium. The entire structure together with asci looks like a compact hollow small pin head. This structure is the fructification or the ascocarp. As the ascocarp has no opening in the peridium (to release the ascospores), it is called the cleistothecium.

The mature cleistothecium is a hollow globose structure, yellowish or brownish in colour. Internal to the multilayered wall are found a number of ascogenous hyphae, asci, etc. Soon these degenerate releasing the ascospores into the cavity of the cleistothecium. The spores are released by the decay of the cleistothecial wall

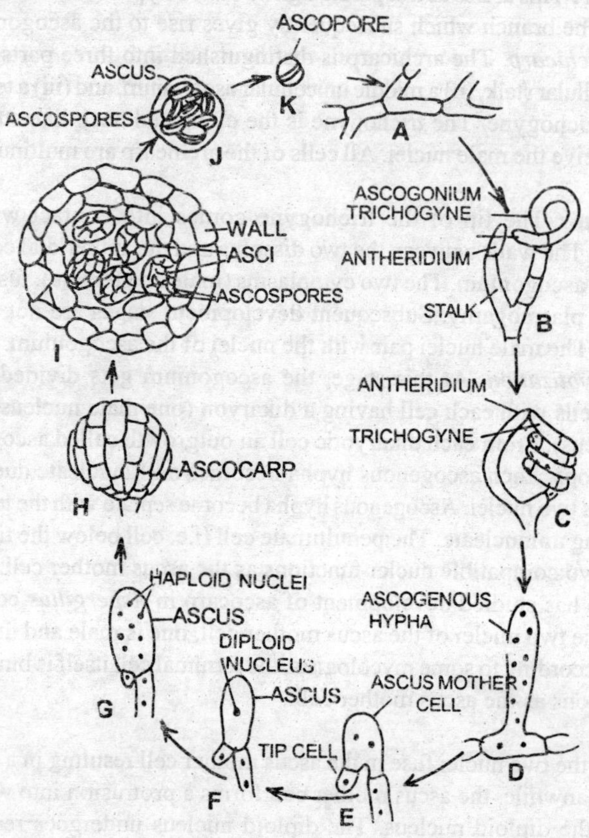

Fig. 11.21 Fungi
Life cycle of Aspergillus

Structure of the ascospore : Each ascospore is multinucleate, and pully wheel shaped. The spore wall is two layered. The outer one is the thick and highly sculptured epispore, and the inner one is the thin endospore. On the availability of a suitable medium, each ascospore germinates into a new mycelium

Laboratory culture of Aspergillus :

A moist piece of bread kept under a bell jar shows the growth of greenish blue cottony mycelium within a few days.

Economic importance

Aspergillus has both harmful and beneficial activities from the point of view of human beings.

armful activities

1. Many species of *Aspergillus* such as *A.glqacus,* and *A.repens* are responsible for the spoilage of exposed food stuff like jams, jellies, bread, leather, textiles, tobacco and many other poroducts.
2. Many of the species are pathogenic to animals as well as human beings. *A.flavus, A.fumigatus,* and *A.niger* cause disease of respiratory tracts commonly referred to as Aspergilloses (singular asprgillosis). Aspergillosis is reported in birds, cattle, sheepl, horses and human beings. Symptoms of aspergillosis resemble those of tuberculosis.
3. Disease of the human ear called *otomycosis* is caused by *A.niger, A.flavus* and *A.fumigatus.*
4. *A.flavus* produces a toxic substance called afflotoxin, which has some carcinogenic effects and may cause cancer of a liver in human beings and animals.
5. Conidia of Aspergillus are abundant in air. They usually spoil the laboratory cultures.
6. Many plant diseases – Crown rot of ground nut ball and rot of cotton are caused by species of *Aspergillus.*

Useful activities

1. Several species are employed in cheese manufacture.
2. *A. oryzae* is used in the preparation of wine from rice and soyabean sauces.
3. *A. niger* is used in bio assay of metals, as it can detect copper even in traces.
4. Some species of *Aspergillus* are the source of certain antibiotics like Flavicein, Aspergillin, Geodin, Fungalin, Patulin Ustin etc.
5. *A. gowssipii* is used in the production of vitamin B.
6. Some species are used in the reproduction of fats.

7. Several species are used in the Industrial production of organic acids like citric acid and gluconic acid.

CLAVICIPITALES

The order comprises of many economically important members, having pseudoparenchymatous structures called stromata. The prominent numbers of the order are *Claviceps* and *Cordyceps,* included in a single family clavicipitaceae.

ASCOMYCETES
CLAVICIPITALES
CLAVICIPITACEAE
CLAVICEPS

Occurrence

The genus *Claviceps* is a parasite which generally infects the members of the family Graminae and causes a disease known as ergot. There are nearly 12 species in the genus, most of which bring about the ergot disease of graminaceious plants. Some of the important species are *Claviceps purpurea, C. microcephala* etc. *Claviceps purpurea,* causing ergot disease of rye plant is most important economically. The life history of this species was established in 1863 by Kuhn. In the ergot disease, the ovary of the host tissue is replaced by a compact mass of the fungal tissue called sclerotium.

Sytematic position

There is no unanimity among mycologists as to the systematic position of *Claviceps.* It is included under Sphaeriales (Miller, 1949), Hypocreales (Martin, 1950), and Clavicipitales (Smith, 1955). In this book Smith's classification is followed.

Life history

The life history of *claviceps* comprises of three main stages which can be clearly distinguished. These different stages are :

(1) The *Sphacelia stage* or *Honey dew stage*
(2) The *Sclerotium stage*
(3) The *Ascostroma stage*

Sphacelia stage

Sphacelia stage is found in various grass plants. In the rye plant, this stage is initiated by the ascospores.

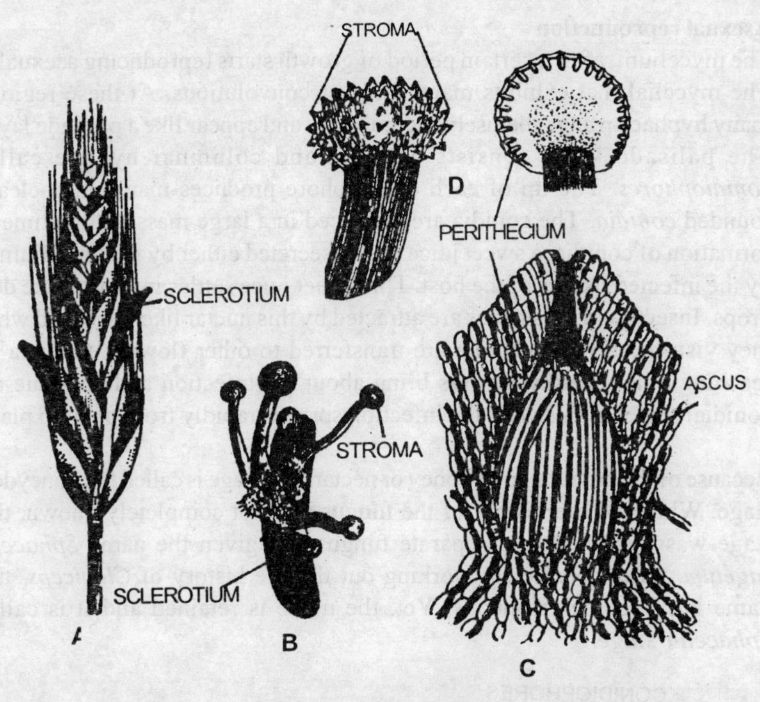

Fig. 11.22 Fungi

Claviceps. Stages in the life history of Claviceps purpurea

A. Infected ear or rye, **B.** Germinating sclerotium, **C.** Long section of ascostroma toshow the Perithecia, **D.** A Perithecium showing asci.

The ascospores of *Claviceps* are long and needle like. They are of light weight and usually wind dispersed. Infection of healthy plants takes place generally during spring season, when the plants are in flowering. The ascospores reach a susceptible host plant and continue the further development. On falling upon a flower, either on the stigmatic surface or on the ovary directly, the ascospores produce germ tubes which by continuous growth reach the ovary. When, once the germ tubes establish themselves in the ovarian bae, further growth takes place and mycelium is established. Initially, the mycelium grows within the tissues of the ovary. But after a certain period, it also comes out of the ovary and envelopes it with the mycelial growth. White, thin, cottony mycelial envelope may easily be observed at this stage when a flower is dissected out.

Asexual reproduction

The mycelium, after a certain period of growth starts reproducing asexually. The mycelial mat exhibits many folds or convolutions. At these regions, many hyphae arrange themselves compactly and appear like a palisade layer. The palisade layer consists of erect and columnar hyphae called *conidiophores.* The tip of each condiophore produces many uninucleate, rounded *conidia.* The conidia are produced in a large mass. At the time of formation of conidia, a sweet juice is also secreted either by the mycelium or by the infected tissues of the host. This sweet juice settles in flower like dew drops. Insects and other flies are attracted by this nuctar like juice, and when they visit the flower, conidia are transferred to other flowers. Conidia on reaching other healthy flowers bring about the infection and continue the conidial stage. In this way, the infection spreads rapidly from plant to plant.

Because of the production of honey or nectar, this stage is called the honeydew stage. When the life history of the fungus was not completely known, this stage was identified as a separate fungus and given the name *Sphacelia sagetum.* After completely working out the life history of *Claviceps,* this name has become erroneous. Yet, the name is retained and it is called *Sphacelia stage.*

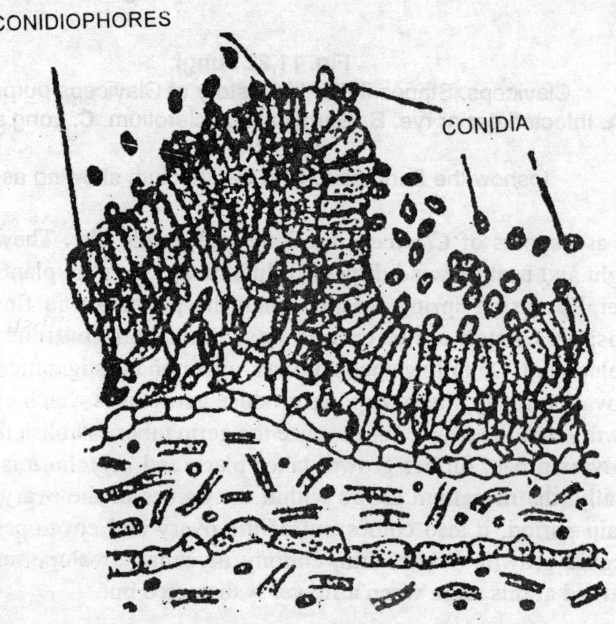

Fig. 11.23 Fungi
Claviceps. Asexual stage showing condiophores and Conidia

Sclerotium stage

The mycelium in the ovarian tissue continues its growth, and eventually occupies and destorys the entire ovary. The ovary is completely replaced by a compact, hard, dark brown mass of the fungal tissue called the *Sclerotium*. The sclerotium may sometimes be pink or purple in color. Occasionally, the sclerotium may be enveloped by the hyphae of the sphacelial stage. In shape, the sclerotium will be like the grain rye but longer than it. It is this sclerotium that constitutes the ergot of commerce. Generally in an ear (inflorescence), all the grains will not be affected by the fungus. Hence, it is not uncommon to see ergots and healthy grains in the same ear.

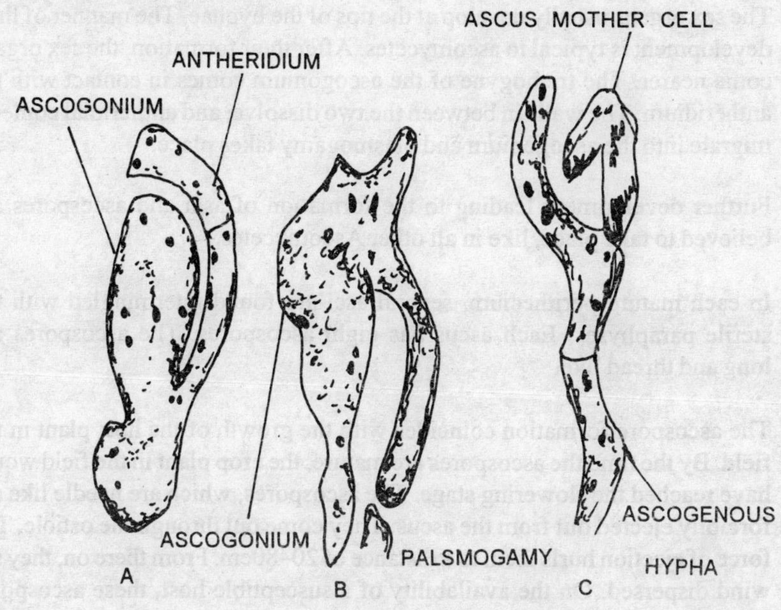

Fig. 11.24 Fungi

Claviceps. Sex organs of C. purpurea

A. Antheridium and Ascogonium, **B.** Plasmogamy, **C.** Formation of Ascus

The sclerotium stage generally constitutes the resting stage of the fungus. The formation of the sclerotium coincides with the maturation of the healthy grains in the host plant. The sclerotium undergoes a period of rest in the winter. During harvest, the sclerotia fall on the soil where they undergo rest. While threshing the grams also, sclerotia get incorporated in the soil. If a new crop is raised in the same field, infection is ready in the soil to attack the plant.

The resting sclerotia come to active life during the next spring. Germination of sclerotium is initiated by the development of 10–12 short stalk like structure. These stalk like structures have a small rounded head. Each one of these outgrowths is called *ascostroma*. The *ascostromata* are about three eights of an inch in length. In a young *ascostroma,* all the hyphae are homogenously arranged. Subsequently, however, small cavities appear in the peripheral region of the stroma. These cavities open to the exterior through small openings called *ostioles*. These cavities are *perithecia*. It is in these perithecia that sex organs are formed.

The sex organs are *antheridia* (male), and *ascogonia* (female). These are formed by the hyphae that constitute the inner lining of the perithecial cavity. The sex organs usually develop at the tips of the hyphae. The manner of their development is typical to ascomycetes. After their formation, the sex organs come nearer. The trichogyne of the ascogonium comes in contact with the antheridium. The walls in between the two dissolve, and antheridial contents migrate into the ascogonium and plasmogamy takes place.

Further development leading to the formation of asci and ascospores are believed to take place, like in all other Ascomycetes.

In each mature perithecium, several asci are found intermingled with the sterile paraphyses. Each ascus has eight ascospores. The ascospores are long and thread like.

The ascospore formation coincides with the growth of the host plant in the field. By the time the ascospores are mature, the crop plant in the field would have reached the flowering stage. The ascospores, which are needle like are forcibily ejected out from the ascus. They come out through the ostiole. The force of ejection hurls them to a distance of 20–80cm. From there on, they are wind dispersed. On the availability of a susceptible host, these ascospores bring about infection and continue the life cycle.

Economic importance

The ergot of rye which is is the sclerotium of the fungus is economically very important. Though very little damage is caused to the crop yield, because not all grains get deformed into ergots, there is a danger of its consumption together with the grains. During the threshing of the crop, the ergots get mixed with grains and if consumed, produce a physiological disease called *Ergotism*. *Ergotism* is produced in both human beings as well as in domesticated animals. In human beings ergotism results in loss of hair, teeth and nails and gangrene. In cattle also, gangrene is seen besides loss of hooves, tails and horns.

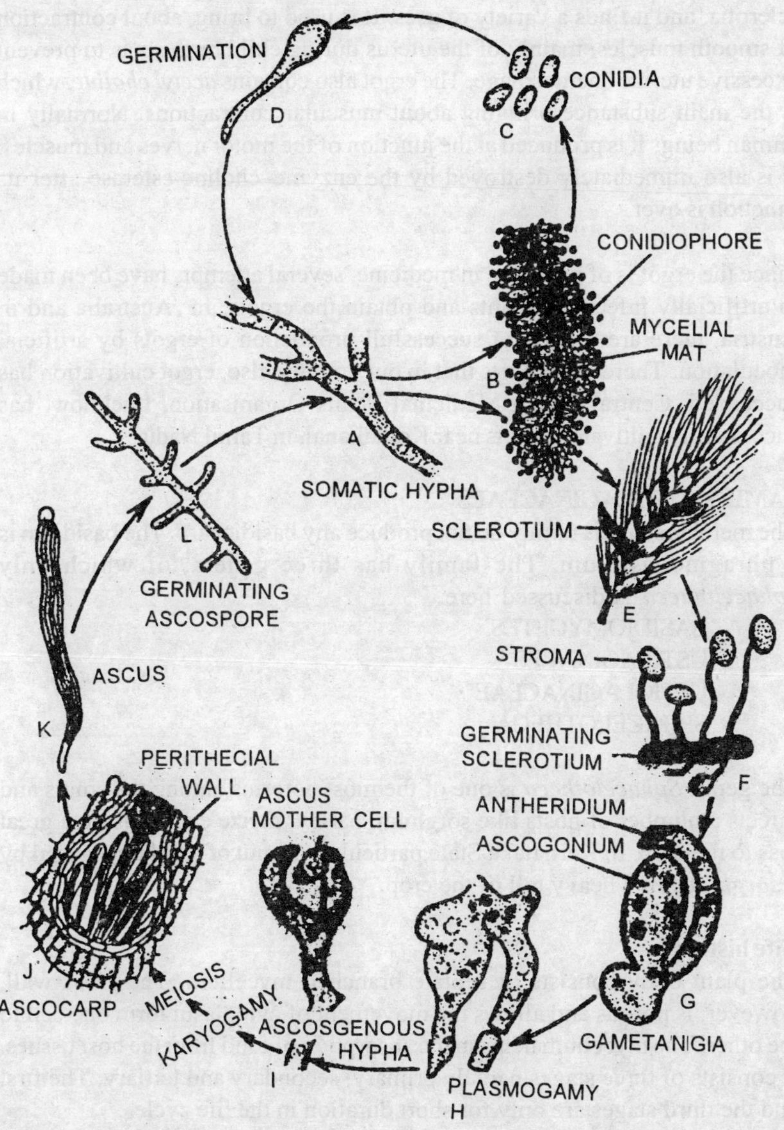

Fig. 11.25 Fungi
Claviceps Pictorial representation of the life cycle Ascostroma stage

The ergot has a positive importance also. It is rich in several alkaloids, chief of which are *ergotomine* and *ergotoxin.* The drug ergot, is obtained from the sclerotia, and it finds a variety of uses. It is used to bring, about contraction of smooth muscles, mainly of the uterus during child birth so as to prevent excessive uterine haemorrhage. The ergot also contains *acetyl choline,* which is the main substance bringing about muscular contractions. Normally in human beings it is produced at the junction of the motor nerves and muscles. It is also immediately destroyed by the enzyme–choline esterase after its function is over.

Since the ergot is of great use in medicine, several attempts have been made to artificially infect rye plants and obtain the ergots. In Australia and in Austria, there are reports of successfull production of ergots by artificial inoculation. There are reports that in our country also, ergot cultivation has succeeded. Central Indian Medicinal Plants Organisation, Lucknow, has successfully cultivated ergots near Kodaikanal, in Tamil Nadu.

FAMILY : USTILAGINACEAES

The members of this family do not produce any basidiocarp. The basidium is a phragmobasidium. The family has three genera, of which only *Sphacelotheca* is discussed here.

BASIDIOMYCETES
USTILAGINALES
USTIGLAGINACEAE
SPHACELOTHECA

The genus *Sphacelotheca* is one of the most notorius among the smuts and infects a number of hosts like sorghum, barley, maize etc., causing a great loss to the crop. In Karnataka State particularly, smut of sorghum caused by *S.sorghii* takes a heavy toll of the crop.

Life history

The plant body consists of septate branched mycelium. The cross wall, however, is porous and allows the movement of cytoplasm form one cell to the other. The mycelium helps in the absorption of food from the host tissues. It consists of three stages namely primary, secondary and tertiary. The first and the third stages are only for short duration in the life cycle.

The primary mycelium in monokaryotic and the nuclei are haploid. A monokaryotic mycelium is produced by the germination of the basidiospore. Soon ,a secondary mycelium is formed from the primary mycelium as a result of sexual reproduction. Since only plasmogamy takes place, the cells of the secondary mycelium will be dikaryotic (sexual reproduction is somatogamous).

The symptoms of the infection are manifested only by the secondary mycelium. Primary mycelium, inspite, of its presence in the host tissues does not cause much damage, and also does not show any morphological symptoms.

Usually, when an infection is brought about by the basidiospore, it first produces the primary mycelium. The basidiospore gains entrance either through the stomata or may fall on the stigma of the flower and enter into the ovary directly.

The disease symptoms are usually mainfested only in the inflorescences. No symptoms are seen in the vegetative parts. The secondary mycelium whether formed directly in the ovary or in the vegetative parts, starts reproducing only in the ovaries.

Reproduction of the secondary mycelium :
The cells of the secondary mycelium isolate themselves, acquire thick walls and transform themselves into chlamydospores (brand spores or smut spores or telutospores).

The teleutospores are unicellular and dikaryotic. They are usually spherical in shape and have a thick two layered wall. The outer layer, which may be smooth or spinescent, is called the endosporium.

The chlamydospores are produced in large numbers inside the ovary. Due to the infection, each grain will turn into a compact mass of fungal tissue called ths sorus. The sori may be long or short. The tough wall of the sorus may entirely be composed of the fungal tissue or may also have the host tissue. Inside the sorus is a central hard columella extending from the base to the apex. This represents the remnants of the host tissue, and is composed of parenchyma and vascular elements as in *S.sorghi.* Between the walls of the sorus and columella, the spore mass is held. The spores are usually round or even oval; olive brown in colour and measure about 5–9 μ in diameter.

Depending on the time of liberation of the spores, the smuts may be classified into two types. *Loose smut* and *Covered smut.* In loose smuts, the spores are discharged directly from the grain and they are naked. Whereas, in covered smuts, the spores are covered by means of a membranous sac and they are liberated frequently during the threshing of the grain. Eg: Loose smut of sorghum is caused by *S. cruenta.*

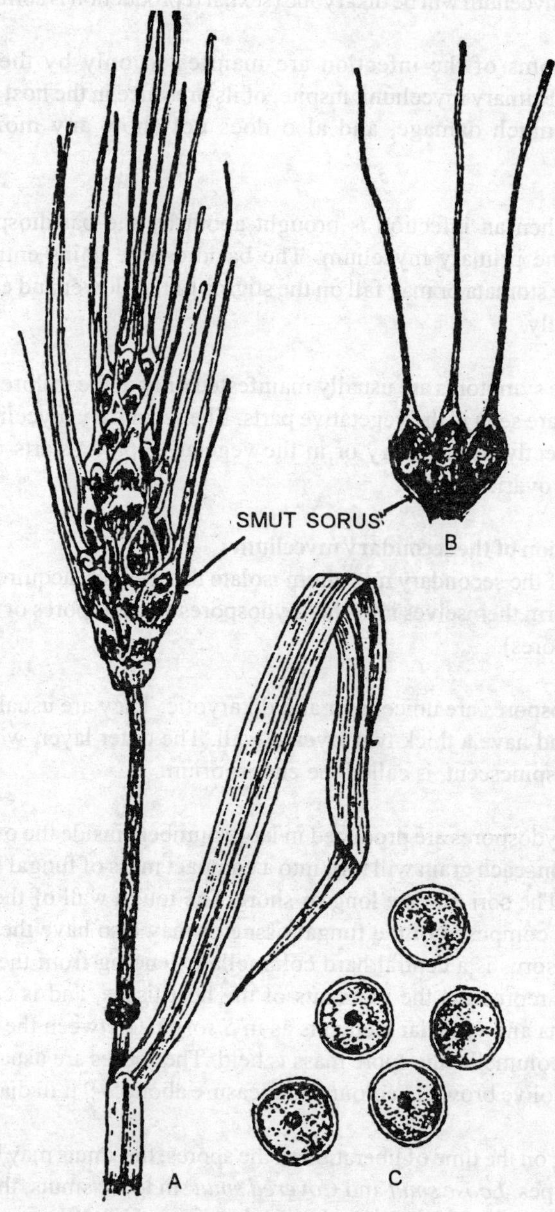

Fig. 11.26 Fungi
A. Infected ear, **B.** Smut sorus, **C.** Chlamydospores

Germination of the teleutospores (Chlamydospores)

The teleutospores germinate immediately either in the soil or on the surfaces of infected seeds, or on the stigma of the ovaries. Usually, a temperature range of 28–38°C will hasten the process of germination. But, the spores are viable for several months as has been proved by experiments.

At the beginning of germination, karyogamy takes place in the teleutospore forming a diploid nucleus. The exospore ruptures and the endosporium comes out in the form of a hyphal tube callled the promycelium or the basidium. The diploid nucleus migrates to the basidium where it undergoes reduction division into 4 uninucleate cells. Each cell produces a sterigma at the tip of which, a basidiospore is formed. The liberated basidiospores start the life cycle. Some times, the entire promycelium is disjuncted from the spore and in some cases branching hyphae develop in place of basidia.

Physiological specialization

The different species of *Sphacelotheca* are known to process several physiological races. The sori produced by these races exhibit definite differences in their color, length and manner of rupturing. Five races are reported to occur in the United State in *S.sorghii.* Vaheuddin (1942), has reported several physiological races of *S. sorghii* from India.

Physiological specialisation may be defined as the existence of several morphologically identical individuals differing in their physiology, and specialisation to a particular host. Physiological specialisation is mostly due to genetic characters.

Economic Importance

The genus, *Sphacelotheca,* causes several disease to many of our agricultural crops. Some of the diseases are given below.

Name of the disease	Casual organism
Grain smut of Sorghum	*Sphacelotheca sorghii*
Loose smut of Sorghum	*S.cruenta*
Head smut of Sorghum and maize. *S. reliana*	

Grain smut of sorghum

Grain Smut of Sorghum caused by *S.sorghii* is the commonest and the most destructive disease of sorghum in India. The smut infects exclusively the ovaries, and all the grains in a ear turn into smut sori. Infected ovaries turn into greyish spore sacs enclosed by a wall made up of fungal hyphae. Each sorus has a central columella made up of a hard tissue surrounding which are found a central of spores. The sori measure in length from 1/6" to ½". Within the spore sac, sterile cells are also found intermixed with the spores. Spores are globose to oval, dark brown in colour, smooth walled and 5–9 μ in diamtere. Spores readily germinate on moist soil and remain viable for more than 5 years.

The smut is externally seed borne, and the spores infect the host between the period of germination and the emergence of the seedling above the ground level. Due to the infection, the metabolism of the host is disturbed and it forms fewer nodes than does the normal plant.

Control measures

Since the disease is externally seed borne, seed treatment or dressing with fungicides yields a good result. Organic mercury compounds and certain non-metallic organic compounds have proved to be very good in controlling the disease. Seeds can be treated with 0.5% formaline for two hours, or with 0.5 to 3% copper sulphate for 3–15 minutes. Solar energy treatment for four hours is also good in summer months. In our country, fine sulphur dust isused extensively because of its ease in application and also of its cheap cost. The use of clean seeds free from smut sori, is of course an important precaution.

Loose smut of sorghum

Loose smut of sorghum in not as common as the grain smut in our country. It resembles the grain smut very closely. It occurs in our country in the states of Andhra Pradesh, Maharashtra, Karnataka and Tamilnadu.

The infected plants are shorter with thin stalks, and the ears are produced earlier than in the healthy plants. All the spikelets of the ear get the disease. The covering membrane of the sorus ruptures early releasing the powdery mass of dark spores. The columella in the spore sac persists even after the dissemination of the spores. The disease is caused by *S.cruenta,* (kuhn) Potter. The wall of the sorus is made up loosely joined fungal cells. The spores are round or elliptical, darker brown than in *S.sorghii,* with echinulate wall and 5–10 μ in diameter. On germination, they produce promycelium. Germination can occur at a temperature ranging form 8–30°C. This can be easily cultured on PDA or Czapek's agar media. It can hybridize with *S.sorghi* and *S.reliana.*

Spores retain viability for nearly four years. Low temperature, low soil moisture and deep sowing favour infection. Infection occurs at the time of germination.

Control measures

The methods recommended for the treatment of grain smut are effective for this disease also. Where spores can survive in soil, crop rotation and field sanitation are advised.

Heat smut of sorghum and Maize

This is found mostly in Kashmir causing significant damage to the maize crop. However, the disease has also been reported from Tamilnadu, Karnatka Maharashtra, Madhya Pradesh, Uttar Pradesh, Punjab and Haryana. In the infected plant, the ear is completely damaged. In maize, large sori of the smut replace the tassel. In such cases, floral bracts grow out into leafy structures; sometimes into small shoots. In sorghum, the entire inflorescence is converted into a big sorus about 4" long and 2" wide. At the early stages, a thin white membrane encloses the sorus. A network of black fibres traverses the spore mass and remains adhering, even after the liberation of the spores.

Head smut is caused by *S. reliana* (khun) Clinton. Spores are reddish brown to black, finely echinate, irregular to spherical in shape and 9–12 μ in diameter. Two physiological races exist in *S. reliana.* Spores retain viability for 2 years. Though the disease is externally seed borne, the major source of infection is the soil borne inoculum. Only the young plants are susceptible to the infection, low temperature favouring it.

Control measures

Effective control can be brought abut by a co-ordinated approach comprising field sanitation, croprotation, and seed treatment. Disinfection of the seed is ineffective in South India. Since the disease spreads slowly, destroying of infected plants will save the other plants from getting the infection.

DEUTEROMYCETES
MONILIALES
TUBERCULARIACEAE
FUSARIUM

Occurrence : The genus *Fusarium,* is one of the largest in the family. It occurs as a facultative parasite and common in the soil as a saprophyte. Some species (soil inhabitants), are capable of living indefinitelyin the soil as saprophytes, while others (soil invaders) soon die out in the absence of a proper host plant. Many important plant diseases such as damping off of seedlings, 'root rot' and wit diseases are caused by *Fusarium.* Some of the diseases caused by *Fusarium* are listed below –

1. *F. udum* – Wilt of Arhar *(Cojanus cajana)*
2. *F.lini* – Wilt of flax *(Linum usitatissimum)*
3. *F. orthaceras* – Wit of gram *(Ciceer arietinum)*
4. *F.vasinfectum* – Wilt of cotton *(Gossypium spp)*

5. *F. lycopersici* – Wilt of tomato *(Lycopersicon esculentum)*
6. *F. cubense* – Wilt of banana *(Musa spp)*

In wilt diseases, the mycelium invades the vascular tissue and finally blocks the xylem vessels. *Fusarium* is also known to produce toxic secretions which block the vessels.

Somatic structure : The mycelium has much branched, separate and multinucleate hyphae. The hyphae (when present inside the host), are restricted to the vascular tissues and are both intercellular and intracellular. The hyphae are hyaline in the beginning but older ones are dark in colour. The hyphae form matted coils within the vascular tissues of the root and at the basal portion of the stem, thus resulting in the plugging of the lumen of the xylem vessels. The blockage of the lumen interferes with the free flow of water movement in plants which consequently wilt.

The hyphae are also known to secrete Fusaric acid, which is toxic to the xylem vessels.

Asexual reproduction : Asexual reproduction is brought about by means of three types of spores viz.., *macroconidia, microconidia* and *chlamydospores.* These are produced by the mycelia within the host plant.

Macroconidia are formed externally in small cushions of stromatic mycelium called *sporodochia.* These (macroconidia) are separate sickle shaped bodies pointed at both the ends. The macroconidia are borne on short conidiophores, and are shed at maturity without being held together.

Microconidia are well, small, oval, spherical or crescent shaped spores. They may be aseptate or may have one or two septa. They are produced within the host tissues on branched conidiophores. Microconidia may be spherical or oval in shape, and may be produced on small phialides instead of conidiophores. As the microconidial stage of the fungus resembles those of another fungus *Cephalosporium,* it is also often called *cephalosporium stage.*

Chlamydospores are also formed within the host tissues. They are round or oval and thick walled spores, formed singly or in chains of two or more. The spores may be terminal or intercalary on a hypha and remain viable for a long time.

At the appropriate time, conidia (macro and micro) germinate on a suitable substratum into a new mycelium.

Fusarium, being a member of Deuteromycetes, does not show the perfect stage (sexual stage).

Disease symptoms

Fusarium infects the young seedlings which are 5–6 weeks old when the symptoms appear. The wilt is characterized by sudden yellowing or wilting or drying of the leaves. Examination of the wilted plants shows that the roots and the base of the stem reveal the blackenig of the parts. Sometimes small black streaks also appear on the stem.

Control measures

1. Crop rotation.
2. Improving the field sanitation conditions and use of soil disinfectants.
3. Use of resistant varieties.

12

PROTOZOA

Protozoa (*Proto*=first +*Zoon*=animal) are the most privmitive among the living beings that are traditionally classified as animals. Currently, however, they are included under eukaryotic protists which includes many non-animals also.

Protozoans are generally regarded as unicellular organisms. But according to many zoologists, it is not correct to designate protozoans as unicellular as they are complete organisms and not loose cells moving around. The body construction of some protozoans may be more complicated than some of the simple metzoa. Hence, protozoans are to be regarded as–*non cellular* or *acellular* animals.

The protozoans may be distinguished from other eukaryotic protists by their lack of cell wall and locomotion that is seen atleast in some stage of the life-cycle; if not always. The protozoa consist of some 50,000 (some estimate it to be 65,000) organisms, whose cell is bound only by a membrane. Many zoologists regard protozoa as a heterogeneous asemblage of animals and the acellular organization seems to be the only common character for all of them.

Protozoans were first observed by Leeuwenhoek (1671). The other prominent protozoologists who have contributed towards the understanding of these unique living beings are –Jobloot (1718); Pasteur (1870); Butschili (1881); Ronald Ross (1898) etc., The term protozoa was first used by Goldfuss (1817).

As has been mentioned already, protozoa in the present day usage does not indicate a taxonomic group or phylum, even though it was no early. Of the estimated 65,000 described species which can be regarded as protozoans, more than 50 percent are fossils. Of the remaining 50% (about 35,000 species), some 22,000 are free living while 10,000 are parasitic. Among the 10,000 parasitic protozoans only a few are human pathogens, but this itself is causing enough havoc in terms of human health. The notorious malarial parasite which was once thought to have been conquered, is raising its head causing alarm as the parasites are becoming resistant to traditional anti-malarial drugs (quinine).

Occurrence

Protozoa prefer a moist habitat. They are ubiquitous and are found wherever nature offers a damp surface. They are found in sea water, fresh water and also in moist soil. Some of the free living protozoa have even been coloured from the icy Polar regions. Many protozoa also occur in association with other animals. The relationship may be mutualism, commensalism or parasitism. Many of the protozoans are adapted to dry conditions also, which they withstand by undergoing encystment; returning to the active state during favourable conditions.

From the standpoint of ecolgical habit, protozoans are of two kinds, free living and symbolic (including parasitism). Free living protozoa are found in a variety of moist habitats. The chief factors which govern the distribution of protozoa are moisture content, temperature, light, availability, and type of nutrients etc.

Many of the free living protozoans which are chlorophyllous *(Euglend),* require sunlight for their nutrition.

pH of the medium has a bearing on the distribution of protozoa. Although protozoans can survive in the range of 3.2–8, a range of 6.0–8.0 is regarded as optimum.

A medium rich in inorganic and organic nutrients is ideal for the growth of protozoa. Some protozoa thrive very well in water low in organic matter but rich in oxygen content. Others, may prefer a reverse situation. Some of the ciliate protozoa like *Paramecium,* prefer a medium rich in bacteria and other tiny protozoans.

Protozoa prefer a temperature range of 16–25^{0}C although they can survive upto 40^{0}C. Occasionally even warm water springs (30–56^0) are known to harbour protozoans.

Protozoa as a link in the food chain

Protozoans, particularly the marine inhabiting ones, constitute an important level in the trophic organization of the marine ecosystem. In marine waters, *Zooplanktons* (free floating animal like individuals), feed on the *phytoplanktons* (free floating plant like individuals). Most of these zooplanktons are protozoans. They in turn are consumed by other larger organisms like larvae, crustaceans, fish etc.

The food chain starting with phytoplanktons may be represented as follows :

Solar energy → Phytoplanktons → Zooplanktons
(producers) (Consumers)
→ Higher animals. (Secondary and tertiary consumers)

Morphology and cell structure

Most of the protozoans are microscopic rarely reaching a size of more than a few microns. Some of the fossil members (belonging to *Foraminiferida*), however had reached a size of 15cm. *Anaplasma,* a blood parasite is so small that it occupies 1/10 of the space in a red blood cell. *Leishmania donovani,* the human pathogen causing *kala azar* measures 1–4 μm in length. *Amoebaproteus,* may reach 600 μm. Some of the ciliates are very large (comparatively) *Paramecium* is about 2mm; *Spirostamum* is about 3mm long.

The body symmetry in protozoa is non-radial, spherical or bilateral. The body is either naked or covered by pellicle. Sometimes, an exoskeleton may also be present.

The cells or protozoa have an outer boundary cell membrane enclosing cytoplasm and other cell organells like, nucleus vacuoles, golgi bodies, mithocondria etc. In some Protozoa (*Euglena*), plastids are also present.

Cell membrane

This is the outer most boundary enveloping the cell. It is thin, elastic and permeable. It has the structure of a unit membrane and is also known as *plasmalemma.* The cell membrane not only provides rigidity and shape to the cell, but being selectively permeable also regulates the movement of substances from and into the cell. The cell membrane also has sites or receptors to receive mechanical and chemical stimuli.

The cell membrane has a tripartite arrangement as revealed by electron microscope. There is a central electron thin layer surrounded on either side by electron dense layers. The membrane is about 75 A^0 thick, with each layer ranging between 20–30 A^0 in thickness. Chemically, the membrane is lipoproteinaceous and accounts for the permeability of the membrane.

Some unit membranes are also present inside the cell. For e.g. endoplasmic reticulum, nuclear envelop etc.

In some protozoa, in addition to the cell membrane, a compound envelop of a modified structure is present. Called *Pellicle,* this envelop helps in protection, support and movement which the normal cell membrane cannot afford. The pellicle is made up of a continuous layer of filamentous molecules as in some amoebae. In *Euglena,* the pellicle is longitudinally striated and ensures flexibility while in *Paramecium* it is quite rigid. In *Paramecium,* the pellicle has three membranes in which the outer one is fashioned into an array of polygons.

Some protozoa have additional protective coverings external to the pellicle, which adds to their diversity. Known as the cae, shells, tests of loricate, these envelops offer protection. Small pores present in these envelops help in communication.

Cytoplasm

Internal to the membrane is the cytoplasm. Though it is a homogenous matrix in some, in the majority of protozoans, two zoanes are distinguishable in the cytoplasm. These are the ectoplasm and the endoplasm. The ectoplasm is more a jelly like substance, while the endoplasm is some what thin. Cytoplasmic organelles are found predominantly in endoplasm.

Cytoplasm of protozoa encloses a membrane system as in other eukaryotes. This is the endoplasmic reticulum (ER), made up of canals and lacunae. The ER is covered by ribosomes. In addition to ER, the cytoplasm has golgi bodies, mitochondria, plastids, vacuoles and nucleus.

Golgi bodies or dictyosomes, are membranous structures associated with the endoplasmic reticulum and are in the form of vesicles or cisternae. They are believed to be involved in the storage of products of cell synthesis. Dictyosomes are numerous in flagellated protozoans, while in ciliates they are poorly developed.

Mitochondria are found distributed in the cytoplasm. They have the envelop of a plasma membrane with the inner layer forming the cristae.

Plastids are found in flagellated protozoans, such as *Euglena.* Each plastid is bound by a double membrane enclosing a number of lamellae. Photosynthetic pigements are concentrated in the lamellae. Pyrenoids are often found in the lamellae.

The vacuoles of protozoans are of two types – contractile vacuoles and food vacuoles. Contractile vacuoles are found in most ciliates, flagellates and amoebae. The number may vary from one to many per cell. These vacuoles are osmorgulatory in function. They collect the waste from cytoplasm, swell in size, the burst releasing the waste to the exterior. Contractile vacuoles may be simple membranous, enclosures, or complex structures having six radiating canals surrounding a central vacuole, as in *Paramecium*.

Food vacuoles are concerned with the ingestion and digestion of food. There are two types of food vacuoles–*phagocytic* and *pinocytic*. Phagocytic vacuoles consist of large food particles, while pinocytic ones have minute food particles in solution form. Food is digested by the activity of the enzymes produced by *lysosomes*. Lysosomes are formed in the region of the golgi bodies.

The cytoplasm of many protozoans also has fibrous structures belonging to two classes. These are microtubules of 20 cm thick, and tiny filaments of 4–10 μm thick. The filaments have globular protein molecules arranged in one or two rows, while microtubules are cylindrical structures having many rows of protein molecules.

Nucleus

All protozoans cells have one, two, or many nuclei. The nucleus is eukaryotic. In Amoeba, there is a single nucleus, while many ciliates have multiple nuclei. In *Paramecium,* where the cell has two nuclei – one is larger than the other. The larger one is called the *macronucleus,* while the smaller one is called the *micronucleus*. The macronucleus regulates the metabolic activities and regeneration, while the micronucleus is involved in reproduction.

Structurally, the nucleus has nuclear membrane, nucleoplasm, nucleolar substance and chromosomes. The number of chromosomes formed during cell division is constant for a given species.

Chemically, the nucleus has nucleoproteins. Both DNA and RNA are found. RNA is concentrated into nucleoli which may be one or two in number. Protozoan nuclei divide by mitosis as well as meiosis.

Location

Protozoans are capable of movement. Movement may be brought about by specialized locomotor organelles such as –pseudopodia, flagella or cilia. A few protozoa lack lacomotor organelles but exhibit gliding movements.

Pseudopodia (false foot), are characteristic of amoebae, where the cytoplasm forms temporary projections in the direction of movement. The pseudopodia also help in the trapping of food.

Flagella are long fibrous extensions of the cell, and are permanent features of the cell. The number of flagella per cell varies. It is one to many. The flagellum has two parts – a central filament (axoneme), and a surrounding contractile sheath.

Cilia are fine, short threads that extend from the body surface. All cilia in a cell may be of the same length or may vary. They are arranged in longitudinal, oblique or spiral rows. Cilia not only help in locomotion but also in driving the food towards the oral grove *(Paramecium)*.
S

Nutrition

Protozoans are autotrophic (holophytic), or heterotrophic (holozoic). Some are parasitic also. In holozoic protozoans many feeding organelles develop in the cell. These are varied. In Amoeba, pseudopodia engulf a food particle and convert it ito a food vacuole. In ciliates, well developed mouths are present. In *Paramecium,* an elaborate food ingesting mechanism is present (see *Paramecium* for details).

REPRODUCTION

Protozoa reproduce asexually as well as sexually, even though the principal means of multiplication is by asexual method.

Asexual reproduction takes place by cell division. This is of two types with respect to the number of daughter cells formed. In *binary fission,* two daughter cells are formed, while in *multiple fission* many daughter cells are formed.

In binary fission, the nucleus divides into two and a constriction appears in the centre dividing the cell into two halves. In protozoans like Ameoba, which have a soft envelop, the cell constricts along the longitudinal plane dividing it into two. In protozoans with rigid envelop, the cytoplasm first projects out through an opening in the envelop and this portion develops an envelop of its own. Only later the nucleus divides.

Ciliates have transverse fission, while flagellates divide longitudinally. In ciliates, where two kinds of nuclei are present, the division process is a bit complicated. In *Paramecium* during binary fission, the diploid micronucleus divides mitocically, while the mcaronucleus divides amitotically.

After the division , the two daughter cells develop the typical organelles characteristic of the parent, even if they do not get them during division.

In *multiple fission,* the nucleus divides repeatedly into a number of units followed by cytoplasmic cleavage resulting in a number of individuals. Multiple fission is infrequent in ciliates, while quite common in sporozoa (*Plasmodium*), rhizopoda (Amoeba) radiolaria, foramnifera etc.

Budding

This is another kind of a sexual increase seen in some protozoa. It is not similar to the porcess of budding seen in fungi like *Saccharomyces.* In protozoa, budding refers to the formation of motile cells or swarmers from the stationary parent cells. Many ciliates exhibit budding. Buds may be formed externally (exogenous), or internally (endogenous).

Sexual Reproduction

Sexual reproduction is of quite frequent occurrence in protozoa. The methods of sexual increase are varied. Syngamy, conjugation and autogamy are some of the methods of sexual reproduction.

In syngamy, the parent cell(s) forms gametes, which may be similar *(isogamy),* or dissimilar *(anisogamy).* When the gametes are dissimilar they are called *microgametes* and *macrogametes.* Microgametes are smaller and motile, while macrogametes are larger and sluggish.
In ciliates like *Parameuium,* sexual reproduction is brought about by means of conjugation. Conjugation is a temporary or partial union of two individuals to exchange nuclear material. After the exchange, the two individuals separate. In some ciliates, a permanent conjugation is also seen where the two individuals fuse permanently.

Regeneration

The ability to produce afresh the lost portions of an organ or a cell is called regeneration. Many of the protozoans, exhibit this capacity. When the cell of the protozoan is cut into two, the nucleated portion can regenerate the other half of the cell, while the non-nucleated portion degnerates. Nucleus is necessary for regeneration.

Classification of Protozoa

The classification of protozoa has undergone a number of changes since the time when protozoa was regarded as a homogenous taxonomic group and accorded the rank of phylum Protozoa. Since then, protozoologists have realized that the protozoans are varied and diverse, and their grouping in one phylum is not justifiable. Modern workers like Kazloff (1972), and Barnes (1980), believe that the so called phylum protozoa can be made into several independent phyla. Horniberg *et al,* however, regard that protozoa can be a single phylum with several subphyla which were earlier regarded as classes.

Sleigh (1973), also considers protozoans to be a single phylum with several sub-phyla below.

The classification of protozoa that is currently used, is the one proposed by the committee on systematics and evolution of the society of protozoologists. Accordingly, protozoa constitute a sub-kingdom under kingdom protista. A few of the schemes of classifications are given below.

Old Classification

Phylum protozoa

Class 1. Flagellata (Mastigophora) e.g. *Euglena.*
2. Sarcodina (Rhizopoda) e.g., *Amoeba*
3. Sporozoa e.g., *Plasmodium*
4. Ciliata (ciliophra)I e.g., *Paramecium*

The classification given above is mainly based on the organs of locomotion. In class 1, flagella are present, in class 2 pseudopodia are present, in class 4 cilia are present, while in class 3 there are no locomotor organs.

Modern classification (Protozoa as a single phylum).

Phylum protozoa has four sub-phyla :

Sub-phylum 1. Sarco mastigophora
2. Sporozoa
3. Cnidospora
4. Ciliophora

Sub-phylum Sarcomastigophora has two super classes – Masstigophora and Opalinata.

Sub-phylum Sporozoa has only one class – Telosporea.

Sub-phylum Cnidospora has two class – Myxosporea and Microsporea.

Sub-phylum Ciliophora has a single Ciliata.

Modern Classification (Protozoa as a sub kingdom).

Sub Kingdom protozoa is divided into seven phyla. These are –

Phylum 1. Sarcomastigophora
2. Labyrinthomorpha
3. Apicomplexa
4. Microspora
5. Acetospora
6. Myxozoa and
7. Ciliophora

Further divisions of these phyla into sub phyla and classes, and their distinguishing features are given in the following table. (Adopted from Pelczar, et al 1986).

Taxonomic Group		Characteristics
Phylum 1.	**Sacromastigophora**	Nucleus of one type; sexuality essentially by syngamy, Locomotor organelles are either flagella or pseudopodia or both.
	Sub-phylum Mastigophora	Trophozoites have one or more flagells. Binary fission takes place by longitudinal fission. Some groups reproduce sexually.
	Class Phytomastigophorea	Flagellates; Plant like in habit; Chromatophores present. Some members are amoeboid. Sexual reproduction well developed e.g., *Euglena, Chlamydomonas*.
	Class Zoomastigophorea	Members parasitic and animal like; flagella present but chromatophores absent, Flagella one to many. Amoeboid forms may lack flagella, e.g., *Leishmania, Trypanosoma, Girardia, Trichomonas*
	Sub-phylum Opalinata	Flagella present all over the body; asexual reproduction by binary fission. Sexual reproduction involves fusion between motile anisogametes. All members are parasitic, e.g., Opalina
	Sub-phylum Sarcodina	Locomotion by pseudopodia which may be distinct or indistinct. Flagella are not present in the adults. A sexual reproduction by fission. Sexual reproduction occurs by flagellate or amoeboid gamets. Members are mostly free living.

Superclass Rhizopoda	Includes naked and testate amoeba, e.g *Amoeba* and *Arcella.* Another group is foramnifera – *Allogromia.* Locomation occurs by pseudopodia.

Taxonomic Group	Characteristics
Superclass Actinopoda	Usually planktonic habit; body spherical with axopodia (with supportive axial filament); body may be naked or covered with chitin, silica or strontium sulphate. Reproduction asexually as well as sexually. Has two classes – Acantharea and Heliozoea.
Phylum II Labyrinthomorpha	Members parasitic on marine plants. Vegetative cells spindle shaped or spherical e.g., *Labrinthula.*
Phylum III Apicomplexa	Organisms are grouped together by an apical complex including a polar ring, rhopteries and micronemes. Polar ring is composed of coiled filaments; Rhopteries have paired tubular organelles and micronemes are made up of elongate, electron dense structures. All members have a spore forming stage in the life cycle and are parasitic, cilia are absent. Reproduction is a sexual as well as a sexual.
Class Sporozoea	Members parasitic. Locomotion of adults by gliding or by undulation of longitudinal ridges. Some groups have flagellate male gamets. Reproduction–sexual and asexual. Oocysts generally enclose infective sporozoites.
Sub-class Gregarinia	Members infect digestive tracts and body cavities of invertebrates.

	Locomotion by body flexion. Trophozoites are large and extracellular.
Sub-class Coccidia	Trophozoites are generally small and present inside the host cell, e.g., *Toxoplasma, Plasmodium.*
Sub-class Piroplasmia	Parasitic on blood cells of vertebrates. Body size small; shapes varied – pyriform, round, rod like or amoeboid. Locomotion by gliding or body flexion. Transmission by ticks e.g., *Babesia*
Phylum IV Microspora	Members are intracellular parasites infecting invertebrates and specially arthropods. Infected host cells become hypertrophied. Spores are small and up to 6 μm in size. Sporoplasm single with simple or complex extrusion apparatus. Spores have polar tube and cap but no polar capsules. E.g., *Nosema.*

Taxonomic Group	Characteristics
Phylum V Acetospora	Spores have one to many sporoplasms. No polar capsules or filaments. All members are parasitic, e.g., *Haplosporidium.*
Phylum VI Myxozoa	Spores have one or more polar capsules and sporoplasms. Members are parasitic. Cysts are formed in the infected organs of vertebrate hosts e.g, *Ceratomyxa.*
Phylum VII Ciliophora	Members are free living, commensales or rarely parasitic, Locmootion is by cilia which may be simple or

	compound (ciliary organelles). Nuclei in the cell are of two types – Macronucleus and micronucleus. Reproduction – Asexual by transverse binary fission. Sexual – by conjugation, endomixis or autogamy.
Class Kinetofragmino phorea	Somatic ciliature are slightly different from oral cilature. Cytosome position varied – apical, sub apical or mid ventral. Cyto pharynx generally very conspicuous. Compound ciliated structures – whether oral or somatic are absent e.g., *Acineta, Didinium, Balantidium.*
Classs Oligohymenophorea	Members have clear cut distinction between oral and somatic ciliature. Oral structures well defined. Position of cytosome – ventral or near the anterior end. Cysts are common. Some members are colonia. e.g. *Tetrahymena, Paramecium, Vorticella.*
Class Polymenophorea	Cytosome has an efficient food directing mechanism through well developed membranelles. These are oriented to the left of the oral cavity and beat in a circular and clockwise fashion. Paroral ciliature present in one or more lines. Somatic ciliature well developed or reduced to cirri at the bottom of buccal cavity. Members are free living occupying a variety of habitats e.g., *Stentor, Euplotes.*

In this book, the following members of protozoa are described in some detail.

Trypanosoma, Trichomonas, Entamoeba. Plasmodium, Balantidium and

Paramecium.

TRYPANOSOMA

Phylum Protozoa	Sub Kingdom Protozoa
Class Flagellata	Phylum Sarcomastigophorea
Order Kinetoplastida	Class Zoomastigophora Order Kinetoplastida

Trypanosomes are flagellated parasites. They occur as parasites in all classes of vertebrates and are pathogenic to human beings and domesticated animals. The chief human pathogenic trypanosomes are *T. gambiense,* and *T. rhodesiense,* causing the charateristic 'African sleeping sickness". Trypanosomes are mostly prevalent in tropical countries.

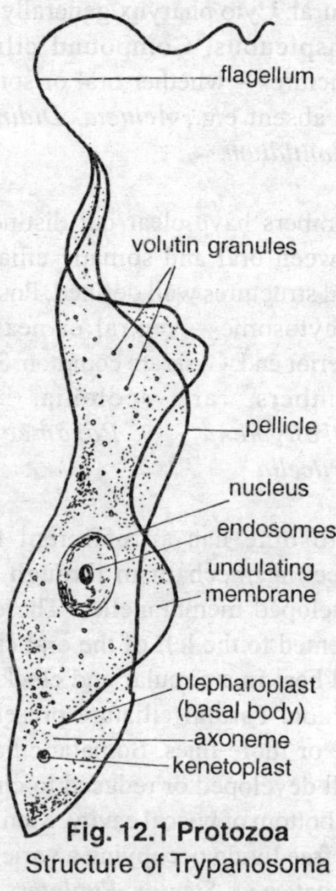

Fig. 12.1 Protozoa
Structure of Trypanosoma

Structure

The body of this protozoan is typically spindle like, with both ends pointed. A firm *pellicle* (membrane complex) surrounds the body and maintains the shape. The pellicle encloses cytoplasm, embeded in which is the nucleus. The nucleus is large and vesicular. There is a single flagellum arising from the posterior end of the body. The flagellum is a long whip like structure arising from a darkly staining granule – the *blepharoplast.* In addition to this, another granule, the parabasal body may also be present. The two structures (blepharoplast and parabasal body), together constitute the kinetoplast. The flagellum runs along the entire length of the body of the edge of an *undulating membrane.* The undulating membrane is an extension of cytoplasm which connnects the flagellum with the body. The flagellum helps in the swimming movement of the parasite in the blood stream.

Nutrition

Food materials are absorbed from the blood plasma by the general surface of the body. The pellicle also helps in gaseous exchange and also diffusion of nitrogenous waste material from the body. There is no contractile vacuole in the cytoplasm.

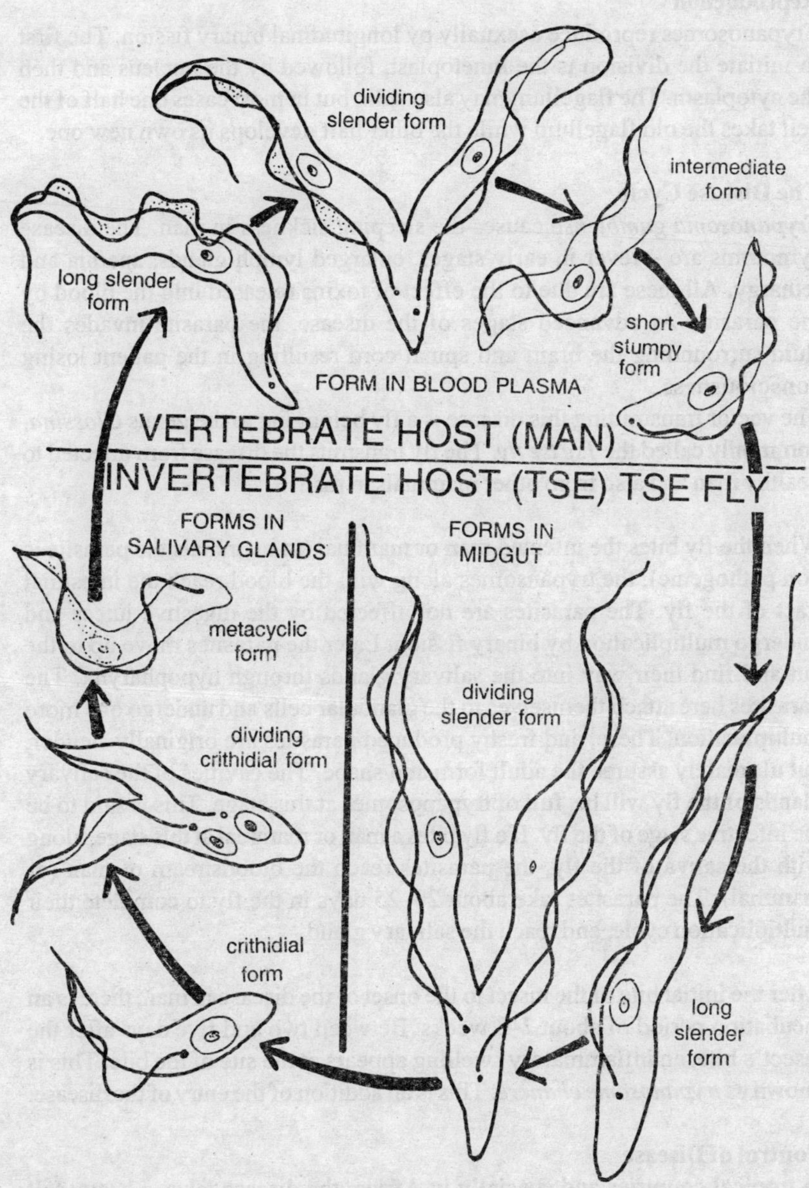

Fig. 12.2 Protozoa
Trypanosoma Life-cycle

Reproduction

Trypanosomes reproduce asexually by longitudinal binary fission. The first to initiate the division is the kinetoplast, followed by the nucleus and then the cytoplasm. The flagellum may also split, but in most cases one half of the cell takes the old flagellum while the other half develops its own new one.

The Disease Cycle

Trypanosoma gambiense causes the sleeping sickness in man. The disease symptoms are – fever in early stages, enlarged lymph glands, anemia and lethargy. All these are due to the effect of toxins released into the blood by the parasite. At advanced stages of the disease, the parasite invades the fluid surrounding the brain and spinal cord resulting in the patient losing consciousness.

The vector transmitting this disease is a fly belonging to the genus *Glossina,* commonly called the *Tse tse fly.* The fly transmits the disease from infected to healthy man and also from other mammals to man.

When the fly bites the infected man or mammal (in mammals the parasite is non pathogenic), the trypansomes along with the blood reach the intestinal tract of the fly. The parasites are not affected by the digestive juices and undergo multiplication by binary fission. Later the parasites move up to the gut and find their way into the salivary glands through hypopharynx. The parasites here attach themselves to the glandular cells and undergo one more multiplication. These, and freshy produced parasites are originally slender, but ultimately assume the adult form and shape. The cavities of the salivary glands of the fly will be full of trypanosomes at this stage. This is said to be the infective stage of the fly. If a fly bites a man or mammal at this stage, along with the saliva of the fly, the parasites reach the bloodstream of man (or mammal). The parasites take about 20–25 days in the fly to complete their multiplication cycle, and reach the salivary gland.

After the initial bite of the insect to the onset of the disease in man, there is an incubation period of about 2–3 weeks. Between two and five days after the insect's bite, an inflammatory swelling appears at the site of the bite. This is known as *trypanosome chancre.* This is an addition of the entry of the disease.

Control of Disease

In tropical countries and specially in Africa, the disease takes a heavy toll and vast tracts are virtually uninhabited due to the ravages caused by the disease. Control measures are very difficult to accomplish because of the large reserviors of trypanosome in wild mammals. Flies can get easily infected by this source. One method of control is eradication of *Tse tse* flies, but it as very difficult to accomplish. Another solution for this problem is to make the toxins of trypanosomes ineffective in man, as they are in wild mammals.

Other Pathogenic Trypanosomes

	Pathogen	Vector	Disease symptoms
1.	*T. cruzi*	Bug *Triatoma megista*	Swelling in the body
2.	*T.brucei*	Glossina spp	Nagana fever in domestic animals
3.	*T.evansi*	Tabanid flies	Surra disease in horses, cattle, camels etc.
4.	*T. Lewisi*	Flea	

TRICHOMONAS

Phylum Protozoa	Sub Kingdom Protozoa
Class Flagellata	Phylum Sarcomastigophora
Order Kinetoplasta	Class Zoomastigophora; order Kinetoplasta

Trichomonas is a common parasite found in many vertebrates including man. Besides mammals, it has also been reported from birds, reptiles, amphibia, molluscs and termites. These species are found in man.

T. homminis – Colon
T. tenax – Mouth
T. vaginalis – Urinogenital tract

Structure : The body of *Trichomonas* is ovoid in shape and somewhat narrower at the hind end. In size, the body ranges from 10–30 μ in length and 5–15μ in breadth. The parasite has four free anterior flagella and one recurved flagellum at the posterior end. The undulating membrane extends halfway along the length of the body to which flagella are attached. The basal bodies are found at the anterior end and these give rise to flagella. Flagella help in the swimming movements of the parasite. There is a single nucleus in the cell.

Electron microscopic studies (Smith and Stewart, 1966), have revealed that the flagellates (including *Trichomonas*), posses longitudinally arranged microtubercules which protrude from the back of the body. Other additional structures are a cytosomal groove, and a parabasal body.

Trichomonas is generally a non-pathogen. Even in human males, where its presence has been recorded from the urethra, apart from causing mild inflammation there are no reports of any complications. In human females, however, *T. vaginalis* causes inflammation and irritation besides leading to the discharge of foul smelling liquid. The parasite lives superficially and there is no record of its invading the tissues. Researches have indicated that hormonal changes result in reduced acidic conditions which might encourage the activity of the parasite. Transmission of pathogen may be *via* male, or due to infected material contaminating the vaginal orifice.

Diagnosis of the disease

Microscopic examination of the fresh vaginal discharge will reveal the presence of parasites in infected individuals. Culture *in vitro* of the vaginal discharge is the most reliable method of isolating and identifying the parasites.

Treatment of the disease

Use of antibiotics have given satisfactory results. Oral and topical use of *Trichomycin* is quite effective. Arsenic based compounds such as *Carabasone,* and the new introduction *Metranidazole,* have proved to be quite successfull in eradicating the infection.

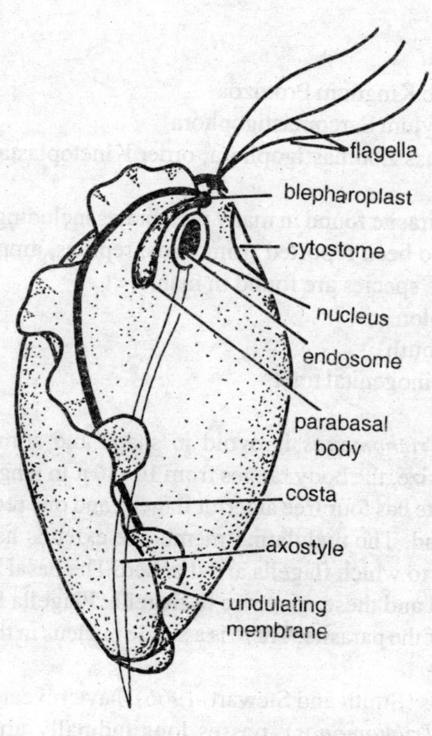

Fig. 12.3 Protozoa
Trichomonas

Most of the amoebae are free living, but some have become endoparasites and have a dapted themselves to live in the intestines of many invertebrates and vertebrates. *Entamoeba* is one such endoparasite, many species of which are pathogenic, some harmless also. *E. histolytica* is the most important among all the members and causes the notorious amoebic dysentry.

ENTAMOEBA

Phylum – Protozoa	SubKingdom – Protozoa
Class – Sarcodina	Phylum – Sarcomastigophora
	Class Rhizopoda.

Historically, *E.histolytica* was first discovered in 1859 by Labie, while its pathogenic nature was confirmed by Losch (1814). Schavdin (1903), gave the name *E. histolytica*.

Distribution

The pathogen is widely distributed being found in tropical, subtropical and temperate regions. Amoebia infection is quite high in India, China and South American countries. According to an estimate, about 10% of human population has amoebic infection (Craig, 1962).

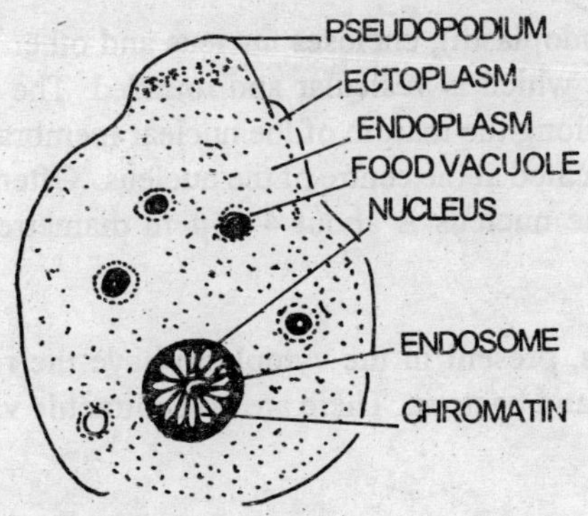

Fig. 12.4 Protozoa
Entamoeba

Habit and Habitat

E. histolytica occurs in a large variety of mammals like dogs cats, rats, monkeys, baboons, chimpanzees, gorillas et.c causing the intestinal infection. In man, it occurs in the mucous and sub mucous layers of the colon. The toxic substance secreted by the parasite dissolves the mucous lining leading to bleeding. From the instestine, the parasite is carried along the blood stream to various organs like liver, kidney, lungs and even the brain. The parasite feeds on the tissues, bacteria, blood cells etc. The disease which is characterized byn loose, blood smeared stools is called amoebic dysentery, which is often fatal if untreated.

Cell structure

The parasite has an active stage called trophozoite apart from the encysted condition.

The trophozoite resembles amoeba in structural details and ranges in size from 15–25 μ. The cell is externally bound by a thin, elastic, selectively permeable membrane (plasmalemma). The membrane encloses the cytoplasm which is divided into an outer ectoplasm and an inner endoplasm. Ectoplasm is thin and hyaline while endoplasm is thick and granular. *Entamoeba* forms a single pseudopodium.

Cytoplasm (endoplasm), encloses nucleus and other organelles. There is a single nucleus which is vesicular and rounded. The chromatin material is concentrated along the margin of the nuclear membrane. The karyosome is small and is located at the centre of the nucleus. Often, it is surrounded by a clear halo. The nucleus is about 4–6 μ in diametre. *Entamoeba* has six chromosomes.

Food vacuoles, present in the cytoplasm have the remnants of the RBC, WBC, tissues and bacteria. There are no contractile vacuoles.

Locomotion

E. histolytica produces only one pseudopodium in the direction in which it wants to move. Hence, it is called monopodial and the pseudopodium itself is called *lobopodium.* In the lobopodium also, there is a clear distinction between ectoplasm and endoplasm.

The movement in *E.histolytica* is a slow forward flowing movement resembling that of garden slug (*Limax spp),* hence, it is called *limax* type movement.

Nutrition

It is holozoic. Fragments of epithelial tissues (that line the intestine), RBCs, bacteria etc. constitute the chief source of nutrition. The parasite is known to produce an exoenzyme – *cytolysin* that dissolves the mucous lining leading to releasing of blood. Blood cells are then trapped by the pseudopodium.

Respiration : *Entameoba* is both aerobic as well as anaerobic.

Excretion : Waste products – mainly nitrogenous, are released in the form of ammonia and diffuse out into the host tissues.

Life Cycle

Reproduction in *E. histolytica* is only asexual. Sexual reproduction has not been reported so far. Asexual reproduction takes place by two methods *binary fission* and *multiple fission.*

Life Cycle includes two stages. Trophozoite stage and cyst stage with a transient precystic stage.

Trophozoite Stage

E.histolytica lives as a harmless commercial in the gut lumen. Ocassionally, for reasons unknown, the trophozoites (the active non cysted, forms), penetrate the mucosa and the muscularis mucosa and invade the sub mucosa. Here they multiply rapidly and spread below the mucosa to form a flask shaped ulcer. Multiplication of parasites is generally by binary fission. The newly formed cells enlarge and continue to attack the host tissues. Some of the parasites, however, under conditions not specifically understood, remain small (7 to 10 μ in diameter) and retreat into the lumen of the intestine. These small forms are called *minuta forms* and are non invasive and non-pathogenic. They eat only the intestinal bacteria. As against this, the larger active ones are called *mangna forms* and are pathogenic. But the current opinion is to regard the so called minute forms as a distinct species. *(E. hartmani)* not belonging to *E.histolytica.*

Encystment

The active aggresively feeding trophozoites migrate to the lumen of the intestine where they undergo encystment. The cells round off and secrete a thick, resistant cyst around them. The cysts of *E.histolytica* are spherical measuring 10–15μ in diameter. The cytoplasm is hyaline and encloses chromatid bodies and glycogen filled food vacuoles. The *chromatid bodies* disappear as the cyst matures. The cyst, however, has a single nucleus at this stage. The nucleus divides twice (not the cell), resulting in the formation of four nuclei. This stage is called the *quadrinucleate stage* of the cyst. At this quadrinucleate stage, the cysts come out of the body of host along with the faecal matter. The cysts are able to live outside the host for many weeks till they reach a new host.

Transmission to new host

The cysts reach a new host along with contaminated food and water. Unclean, unhygenic habits, improper cleaning of food before consuming, use of untreated faeces as fertilizer for crops, improper washing of hands, are the main sources of infection. In addition to the above, some coprophagous insects may carry the cysts on to the uncovered food stuffs. The cysts eventually reach the alimentary canal where they undergo *excystation* (removal of cyst covering), to become active trophozoites once again. In the intestinal lumen, the cyst wall dissolves and a tetranucleate amoeba comes out. This is called the *metacystic* form. This divides to form eight new entamoebae. These newly formed ones migrate into large intestines invade the mucosa, and grow into active trophozoites, and continue the infection cycle all over again.

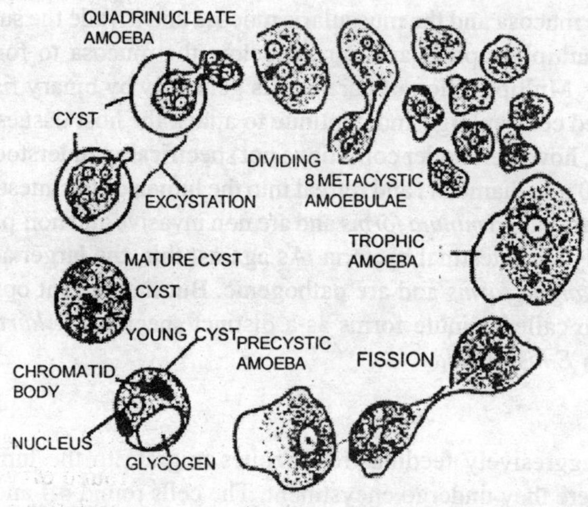

Fig. 12.5 Protozoa
Life cycle of Entamoeba histolytica

Disease symptoms and diagnosis

About 10% of the human population suffer from the infection of the parasite, but most of them are carriers without showing any symptoms. The main symptom of the disease is amoebic dysentery, in which the stools are acidic and contain pure blood (coming out of the intestinal ulcers produced due to the invasion by trophozites)and mucous in which trophozoites are present. Stomach pain is another symptom.

Ocassionally (if untreated), the parasites may invade liver, lungs, gonads, spleen and even brain . They reach these organs along with blood stream. Abcesses of liver, lungs and even brain may be caused during severe infections. Liver damage, brain damage etc., is also not ruled out.

Amoebiasis is usually diagnosed by microscopic observation of parasites in the faeces. In doubtful circumstances, *in vitro* culture may be made.

Treatment

Amoebiasis is difficult to cure. Emetine (an alkaloid) has been quite successful in controlling dysentery. Dehydro emetine, a synthetic derivative is equally effective. Chloroquine, an antimalarial drug cures liver abcesses. In chronic cases, arsenic based compounds like vio form, carbarsone and metranidazole are quite useful.

Prevention

As the proverb goes, prevention is better than cure. The following methods will help in preventing the contact of the disease.

(i) Washing of hand with antiseptic soaps after toilet and before eating food.

(ii) Proper sanitary conditions to dispose faecal matter.

(iii) Sanitary conditions in drinking water.

(iv) Protection of food from insects, flies etc.,

(v) Proper washing of foods and vegetables.

(vi) Use of filtered, boiled water for drinking.

(vii) Avoiding defecating in open areas.

Other species of Entamoeba

Apart from the pathogenic *E.histolylica,* there are other parasitic species also. These, however, are non-pathogenic.

E.coli lives in the colon of human beings and it is estimated to be present in about 50% of human population. It feeds on bacteria, undigested food and other unwanted material.

E.gingivalis lives in the mouth, usually between the teeth and gums. The parasites feed on food particles.

E.hartmani originally thought to be the minute from of *E.histolytica,* lives in the colon. Though it also invades the intestinal tissues, it is less harmful. Trophozoites of *E.hartmani* are much smaller than those of *E.histolytica.*

PLASMODIUM

Phylum Protozoa	Sub Kingdom Protozoa
Class Sporozoa	Phylum Apicompixa Class Sporozoa

Plasmodium, commonly called malaria parasite causes one of the most harmful and as yet unconquered disease of human beings. The disease malaria (*mala;* bad *area* = air), was originally thought to be caused by bad or polluted air. The causative agent of malaria (the parasite) was discovered by Charles Laveran (1880) a doctor in the French army, located at Algiers. He observed the parasite (*Plasmodium*) in the blood of malaria patients. King (1883), discovered that mosquitoes help in the transmission of the disease. Patric Manson suggested that the parasites get into the blood stream of man due to mosquito bite. It was Ronals Ross (1895), who after a detailed study (in 1897), proved that the female *Anopheles* mosquito picks up the parasite and injects it into the human blood. He discovered the sexual cycle of the parasite in mosquito. In his honour, the cycle is called the cycle of Ross. He was awarded the Noble prize in medicine, in 1902, for his work.

There are four different species of *Plasmodium* causing different types of malarial fever. These are :

1. *Plasomdium vivax* – Tertian or benign tertian or vivax malaria.
2. *P.malariae* – Quartan fever
3. *P. ovale* – Oval or mild tertian malaria
4. *P. falciparum* – Subterranean malaria or malignant tertian malaria.

Besides the above mentioned human pathogens, some species, *P. berghei* (in rats), and *P. gallinaceum* (in chicken), are also reported in many animals.

Distribution

Plasmodium vivax is an intracellular parasite seen in the RBCs. It is distributed both in temperate and tropical countries. It requires two hosts to complete the life-cycle. These are

(i) Man or other vertebrates – *Primary host* or *definitive host.*
(ii) Female Anopheles mosquito, or other blood sucking insects – *secondary host* or *intermediate host.*

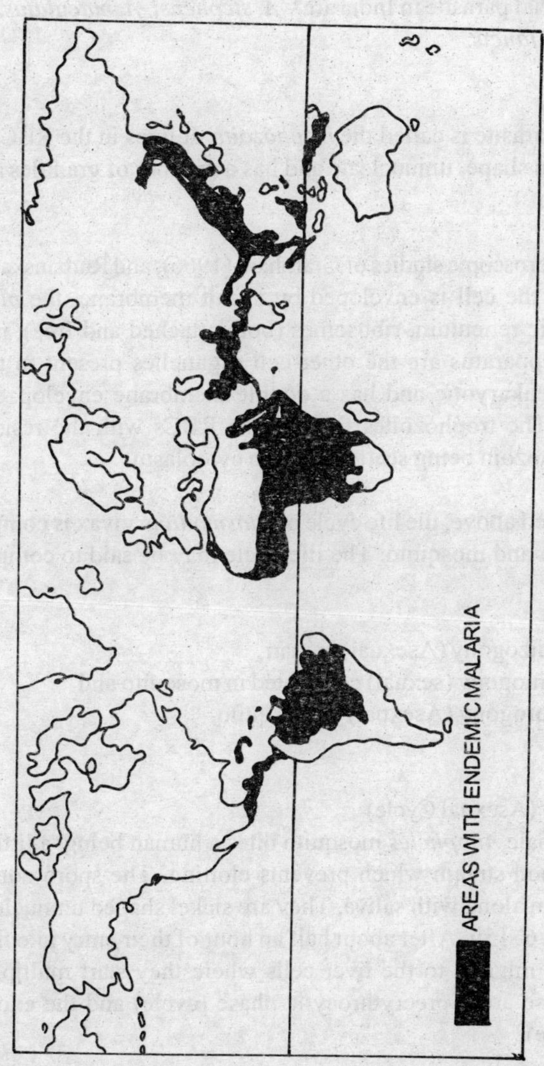

Fig. 12.6 Protozoa
Areas with endemic malaria

A sexual cycle or schizogony takes place in the human beings, while sexual cycle or gamogony and sporogony takes place in the female *Anopheles* mosquito. The species of *Anopheles* which act as vectors for the transmission of the material parasite in India are – *A. stephensi, A. maculatus, A. fluvialitis and A. culcifancis.*

Morphology

The adult parasite is called the *trophozoite.* It lives in the RBC of man. It is amoeboid in shape, uninucleate and has a number of granules and vacuoles in cytoplasm.

Electron microscopic studies of Gramham (1966), and Rudzinska (1969), have shown that the cell is enveloped by a unit membrane, the *plasmalemma.* Endoplasmic reticulum, ribosomes (both attached and free), mitochondria and golgi apparatus are the other cell organelles present in the cell. The nucleus is eukaryotic and has a double membrane envelop. It also has a nucleolus. The trophozoites feed on the RBCs with the residual product called haemozoin being scattered in the cytoplasm.

Life Cycle

As mentioned above, the life cycle of *Plasmodium* vivax is completed in two hosts – man and mosquito. The life cycle may be said to comprise of three stages.

(i) Schizogony (Asexual) in man
(ii) Gamogony (sexual) completed in mosquito and
(iii) Sporogony (Asexual) in mosquito.

Schizogony (Asexual Cycle)

When a female *Anopheles* mosquito bites a human being, a little saliva gets into the blood stream which prevents clotting. The sporo zoites enter the blood stream along with saliva. They are sickel shaped uninucleate, ranging in size from 6–15μ. After about half an hour of their entry into the blood, the sporozoites migrate to the liver cells where they start multiplying in two stages. These are – preerythrocytic phase (cycle) and the exoerythrocytic phase (cycle).

Preerythrocytic cycle

Sporozoites enlarge in size by absorbing nourishment from the liver cells. They (sporozoites) become spherical in shape, and are called *schizonts.* The nucleus of each schizont undergoes multiple divisions to produce a number of daughter nuclei. This multinucleate schizont is called a *cryptozoite.* The cytoplasm of *cryptozoite* undergoes cleavage, and forms a number of uniuucleate cryptomerozoites. The membrane of the cryptozoite breaks open releasing the cryptomerozoites. Eventually, the liver cells also break open and release the cryptomerozoites into the liver sinusoids.

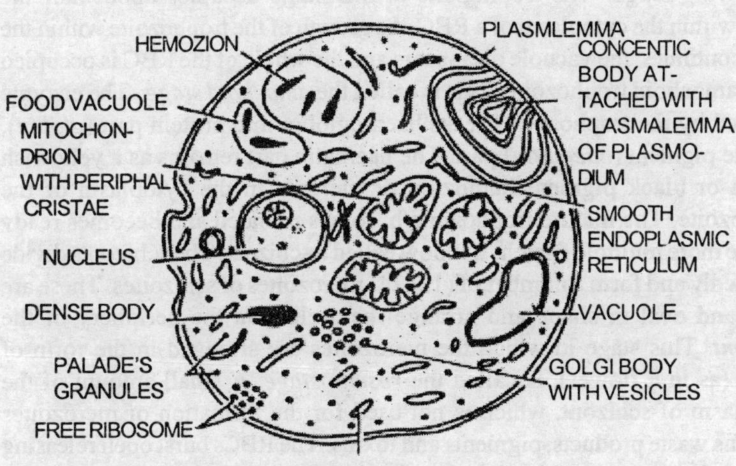

Fig. 12.7 Protozoa

Sporozoite of Plasmodium as seen in electron microscope

Exoerythrocytic Cycle

The cryptomerozoites invade new liver cells and continue their multipli cation and produce adule like cells called *Metacryptomerozoites or phanerozoites.* Two types of metacrypto merozoites are produced- *Micrometa crypto zoites and Macro metacrypto* zoites. Each micrometacryptozoite divides to produce 100–1000 minute cells, which are liberated into the blood stream. Each macrometacryptozoite also divides to produce about 64 cells which invade fresh liver cells and continue the exoerythrocytic cycle. This reservoir of metacrypto zoites helps in the relapse of the fever. As the host is *intracellular*, drugs cannot act on the extra erythrocytic stage of the parasite. The period beginning with the mosquito bite and ending with production of merozoites is called *Pre patent period*. It takes a duration of about ten days.

Erythrocytic cycle or endoerythrocytic cycle

The *crytomerozoites* after their release from liver cells reach the blood, and enter into the RBC and become a spherical trophozoite. Each trophozoite soon develops a vacuole in the centre, pushing the cytoplasm and nucleus to the periphery giving the appearance of a signet ring, hence the name *signet ring stage*. The trophozoite at this stage occupies about half the space within the cytoplasm of a RBC. As growth of the trophozoite within the RBC continues, the vacuole disappears and the whole of the RBC is occupied by an amoeboid trophozoite. This is called the *amoeboid stage*. The enzyme secreted by the trophozoite breaks haemoglobin into protein part (*globin*), and the pigment (*haematin*) part. The haematin part remains as a yellowish brown or black pigment hemozoin or melanin in the cytoplasm of the trophozoite. The mature trophozoite becomes rounded and becomes ready for one more multiple fission. It is now called a schizont. The schizonts divide repeatedly and form a number of (12–24) merozoites or schizoites. These are short and oval in shape and arrange themselves in the periphery of the *schizont*. This stage in which the merozoites are arranged in the form of petals (as in a flower), is called the *rosette stage*. A small amount of the cytoplasm of schizont, which is not used for the formation of merozoites contains waste products, pigments and toxins. The RBCs burst open releasing the merozoites into the blood stream along with the toxins. The toxin initiates the signs of malaria. The merozoites enter new RBCs and start new multiplication cycle releasing more merozoites and toxins into the blood. One crythrocytic cycle, takes about 48 hours to complete. The interval between inoculation of sporozoites into blood, and the first appearance of fever, is called the *incubation period*.

Gamogony

This cycle starts in the man and gets completed in mosquito. After repeated erythrocytic cycles, the merozoites after entering into RBCs transform themselves into spherical *gametocytes* or *gamonts*. The gametocytes do not have a vacuole and possess a large central nucleus and dense cytoplasm. The gametocytes are of two types – *Mycrogametocytes* (male) and *Macrogametocytes* (female). The male gamete is small and has a large centrally placed nucleus, while the female gamete is large and has a small eccentric nucleus. The gametocytes do not divide further but remain inside the RBCs. Further development takes place when these are transferred to mosquito : otherwise they perish.

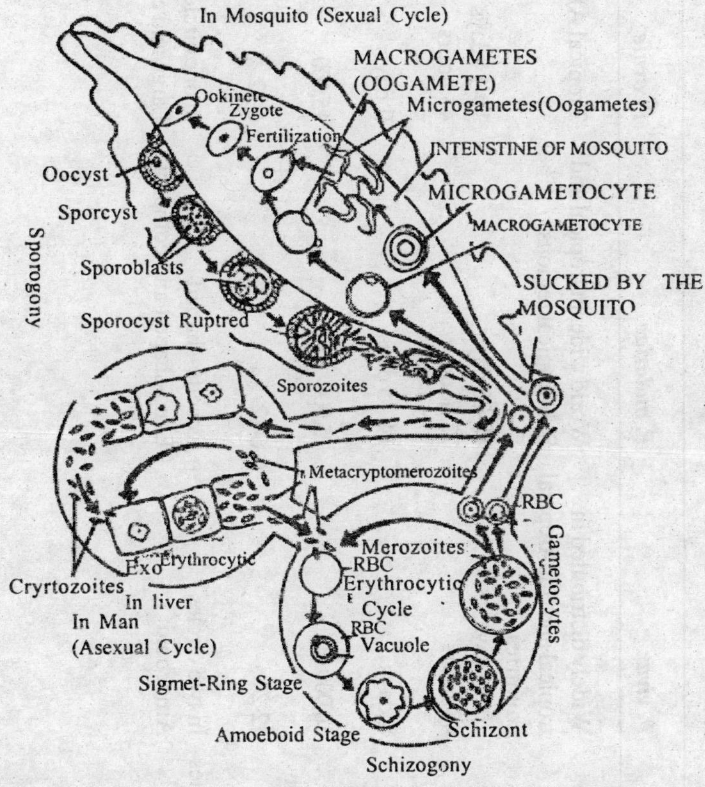

Fig. 12.8 Protozoa
Life-cycle of malarial parasite

A comparative account of four species of human infecting Plasmodium

Character	*P. vivax*	*P. malariae*	*P. ovale*	*P. falciparum*
1. Distribution	Widely distributed in tropical and Sub-tropical countries	World wide, in tropical and Sub-tropical regions	Tropical Africa	Very common is tropical countries
2. Incubation period	10-14 days	18-28 days	10-14 days	9-12 days
3. Duration of erythocytic schizogny	48 hours	72 hours	48 hours	48 hours
4. Duration of exoerythrocytic cycle	8 days	14-15 days	9 days	5 days
5. No. of cryptozoties per schizont	10,000	15,000	15,000	30,000
6. No. of merozoites per schizont	12-24	6-12	6-12	8-20
7. Arrangement of merozoites	In two circles	Rosette-shaped	in one circle	irregular
8. Shape of trophozoites	Amoeboid	Band-shaped and compact	Large and compact	small and compact

9. Infected R.B.C.	Large, distorted and pale	Not enlarged, no dots with Scheffner's dot	Slightly enlarged dots irregular, dots numerous	normal sized, no dots
10. Haematin granules	Light brown, masses	Dark-brown; abundent	Dark-brown in Compact mass	Dark brown in compact mass
11. Schizont in RBC	Occupies more than 2/3 of R.B.C	Almost entire R.B.C.	About 3/4	Less than 2/9
12. Gametocytes	Rounded or oval	Circular or ovoid	Rounded	Creacentric
13. Micro gametocytes	7-8μ wide with numerous male gamets produced	Large nucleus with numerous haemozoin gametes; 4-8 4-8 male gamets produced	Similar to *P. vivax* and scattered bacteria,	Haemozoin golden brown; 2-5 gametes are produced
14. Macrogametocyte	8-10m wide, large few haemozoin granules	Small nucleus, coarse haemozion	Same as in *P. vivas*	Dark, coarse haemozoins around the nucleus
15. Duration of sexual cycle.	10 days	25 days	16 days	10-12 days

Sexual Cycle in mosquito (*The cycle of Ross)*

When a female *Anopheles* mosquito bites a malarial patient, along with blood, merozoites as well as gametocytes (inside the RBCs) reach the alimentary canal of mosquito. There, all the contents except the gametocytes get digested. The gametocytes come out of the RBCs.

The microgametocyte gives rise to 6–8 haploid microgamets which are whip like. They show lashing movement of flagella and shoot out from the surface of the microgametocyte. This process is known as *exflagellation.*

In the macrogametocyte, one or two polar bodies are cut off, before it gives rise to a single haploid *macrogamete or ovum.* The female gamete has a projection called *cone of reception (to receove male gamets).*

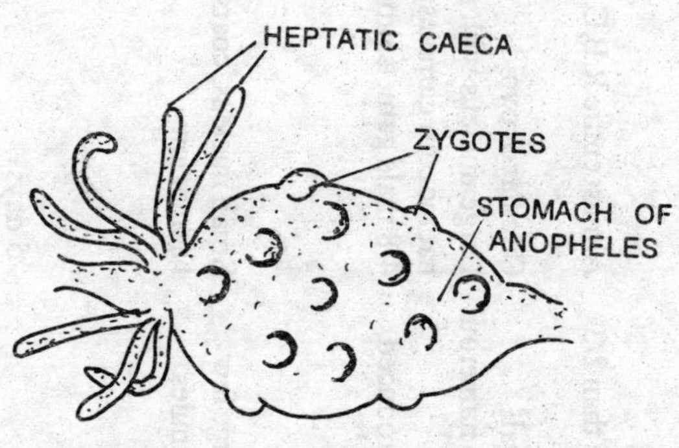

Fig. 12.9 Protozoa
Stomach of infected female mosquito showing oocysts of Plasmodium

Fertilization

The male gametes approach the female gamete and one of them reaches the cone of reception and releases the male nucleus into the ovum. The male and female nuclei fuse, resulting in the formation of a Zygote with a diploid nucleus.

The Zygote remain inactive for sometime but soon become elongated and motile. It is now known as *ookinete* and moves into the stomach of mosquito.

Finally, it reaches the space between spithelial and sub epithelial layer where it undergoes encystment (oocyst). The oocysts grow in size and project into the cavity of stomach from the inner lining. A single mosquito may have 500–5000 oocysts.

Sporogony

This is the post zygotic phase of asexual increase. The nucleolus of each oocyst undergoes repeated divisions (the first one being meiotic), to produce a large number of (10,000) tiny sporozoites. Each sporozoite is a spindle shaped structure. Eventually, the oocyst ruptures and the sporozoites are released into the haemocoel of mosquito. The sporozoites migrate to various organs inside the mosquito, but majority of them reach the salivary glands and remain stationed there. When a mosquito bites a man, along with saliva, the trophozoites also reach the blood stream and start the cycle all over again. It has been estimated that a single mosquito may have as many as 2,00,000 trophozoites.

Pathology of Malaria

Malaria fever is caused due to the release of toxin (haemozoin) into the blood stream once every 48 hours; the fever recurrs every third day. (in *P. vivax).* In *P. malariae,* the fever occurs every 72 hours and in *P. falciparum* and *P. ovale* fever repeats every third day. The following table gives a comparative account of pathogenicity of the four species of *Plasmodium.* (Table adopted from – A text book of Invertebrates – H.S. Bhamrah et al; 1992).

Early symptoms of malaria are nausea, coating on the tongue, headache and muscular pain (Prodromal sympton). Later symptoms (*Paraxym*), occur after a few erthyrocyctic cycles. This includes the following stages.

1. *Rigor stage* : Patient has chill and shivering with rapid breathing and high pulse rate.
2. *Febrile stage* : Shivering subsides and fever rises to about 104ºF to 106ºF.
3. *Defervescent stage* : Profuse sweating; after few hours reduces the temperature.

Control of malaria

During the 1960s it was thought that malaria was completely checked, but it is raising its head again with the parasites developing drug resistance. Malarial control involves a three pronged approach.

1. Eradication of vector (Mosquito).
2. Prevention or prophylaxis and

3. Treatment of the disease.

Eradication of mosquito

It is easier said than done, but the following methods are quite useful in controlling, if not eradicating the disease.

(a) Spraying of DDT, BHC etc.,
(b) Fumigation with pyrethrum, cresol etc.,
(c) Trapping and killing the insects in small boxes.
(d) Sterilizing the males and releasing them in nature where they complete with fertile males.
(e) Stagnant dirty water should not be allowed to settle.
(f) Spreading kerosine or paraffin oil on the surface of stagnant water.
(g) Larvae eating fish or insectivorus plants can be used to eliminate mosquito larvae.

Prophylaxis : (methods to prevent the spread of malaria)

1. Construction of houses at an elevation with full ventilation; with mosquito mesh covering windows.
2. Use of mosquito net during night; use of mosquito repellent mats, coils or creams.

Treatment of the disease

The first drug used to control malaria is *quinine,* obtained from the barks of *Cinchona succirubra.* Other antimalarial drugs are *Chloroquine, Paludrine, Alabrine, Camoquin, Resochin, Pentaquine, Daraprim* etc., Chloroquine and *Daraprim* are supposed to be very effective. A single dose of Daraprim (25mg), eliminates the schizonts of *P.vivax* and *P.falciparum.*

BALANTIDIUM

Phylum Protozoa	Sub Kingdom Protozoa
Class Ciliophora	Phylum Ciliophora
	Class Kinetofragminophorea

The class ciliata has a number of free living members with only *Balantidium coli* as a human pathogenic parasite. It is commonly found in the intestine of man, pig, sheep, camels, cockroaches etc., It is also found in the rectal contents of frog.

B.coli is oval in shape, and may have a size of 200 mm, and is the largest parasitic protozoa found in the intestine of man. The parasites are located mainly in the lumen of the intestine and obtain their food by feeding on bacteria. Occasionally, diarrhoea mixed with blood may result due to *Balantidium infection.*

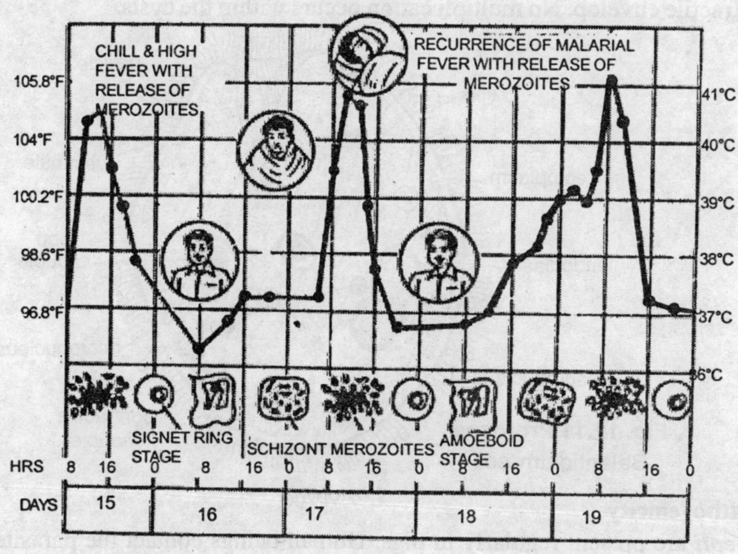

Fig. 12.10 Protozoa
Fever cycle due to infection by P. vivax

The trophozoites are oval in shape, and the entire body is covered with longitudinal and slightly spiral rows of cilia. There is a large prestomial depression or vestibule, leading to cytosome at the anterior end. The broad posterior end bears a cytopyge. The cytoplasm may be distinguished into outer ectoplasm and inner endoplasm. The endoplasm has food vacuoles, two contractile vacuoles and two nuclei – a macronucleus. The micronucleus controls sexual reproduction, while the macronucleus regulates metabolic activities. Nutrition is holozoic with food consisting of tissue fragments, bacteria, erythrocytes etc.

Reproduction in *Balantidium* is both a sexual as well as sexual. A sexual reproduction is by binary fission, while sexual reproduction is by conjugation.

B.coli cysts range in size from 60mm to 70mm. The cyst is covered with a thin refractile envelop. Nó multipli cation occurs within the cysts.

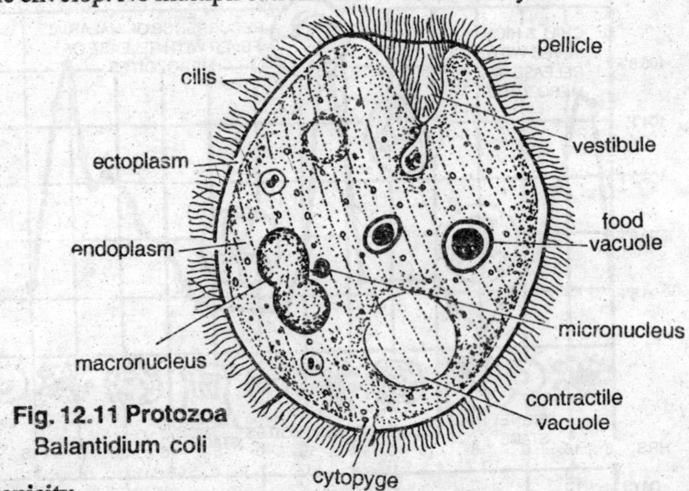

Fig. 12.11 Protozoa
Balantidium coli

Pathogenicity

B.coli are present regularly in pigs. Human beings contact the parasite by consuming water or food contaminated with faeces of swine (pig). The infection is covert and generaly symptomless. When disease becomes demonstrable, the symptoms are–ulceration of colon and diahrroea along with blood.

Treatment

Infection can be avoided by following hygienic methods of food handling and drinking filtered, boiled water. Use of drugs such as oxytetracycline carbarsone or dioiodohydroxyquin can effectively eliminate the infection.

PARAMECIUM

Phylum Protozoa	Sub Kingdom Protozoa
Class Cliophora	Phylum Cilipphora
	Class Kineto frangminoporea

Paramecium belongs to the class Ciliata or Infusoria. The members of this class occur in infusions ie., organic matter decaying in water.

Paramecium is popularly called *'slipper animalcule'* because of the shape of the body. It occurs abundantly in fresh water ponds, which are rich in decaying animal and vegetable remains. A drop of pond water freshy collected will show under the microscope these animalacules moving to and fro in the field (of the microscope).

The body of *Paramecium* is made up of a single cell about 0.3mm in length, and is shaped like the soie of a slipper, hence the name slipper animalcule. One end which is somewhat narrow is called the anterior end, and the broad end is called the posterior end. The anterior end is in front during locomotion. One glide is called the *oral* or *ventral surface* while the opposite end is known as the *aboral* or *dorsal surface.* Present in the *oral surface* near the anterior end is an *oral groove.* The oral groove leads to the mouth (of the animal), called *cytosome.* The mouth opens into a gullet or *cytopharynx* which passes obliquely downwards pointing to the posterior end. Usually, food vacuole is always seen towards the end of the gullet.

Externally, the body is covered by a cell membrane to which are attached a number of small hair like structures called cilia. These are the locomotory organs. The animal moves with the help of cilia and in addition, the cilia also help in gathering food particles due to their lashing movement. The cilia are arranged in longitudinal rows which run lengthwise in the hinderpart of the body, but follow the spiral twist on the oral surface. Two to three rows of cilia lining the gullet are somewhat larger than those present on the general surface. They are fused to form an undulating membrane which is attached to the dorsal wall of the mouth.

Internal to the membrane (*pellicle*) is the cytoplasm divided into two regions– ectoplasm and endoplasm. Ectoplasm is the outer clear non-granular region, while endoplasm is the inner region made up of a granular mass.
The membrane, also called the pellicle is thin, but rigid enough to maintain the shape of the cell. It also affords support to all the internal motions of cytoplasm. The pellicle is made up of a number of hexagonal depressions with a hole in the centre. Cilia emerge out through these holes. These are thin protoplasmic process which have their origin in basal granules. The ciliary movement is regulated by a *neuroneme* or *neuromotor system* which is a sensory fibre running all along the pellicle parallel to it. Present in the dorsal ectoplasm are two contractile vacuoles occupying either end of the cell. Each contractile vacuole consists of a central vesicle and a number of radiating canals which open into it. The whole structure appears star shaped.

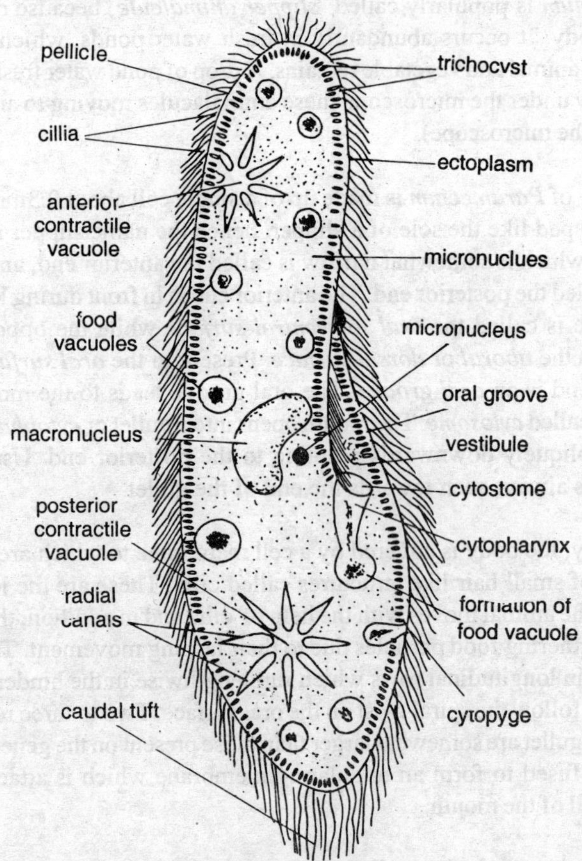

Fig. 12.12 Protozoa
Paramedium, from life

The ectoplasm just below the layer of cilia, consists of a large number of sacs alternating with the basal granules. These sacs called *trichocysts* are spindle shaped, 1/100 mm in size and are filled up with a colourless fluid of high refractive index. When Paramecia are subjected to some external stimulus or irritation, the trichocysts are ejected out through the openings present in the pellicle ridges. During discharge, trichocysts become elongated (about eight times their normal length). The function of these trichocysts are not certain. Some zoologists are of the opinion that they are toxic and serve to paralyze the organisms that they come in contact with. But there is an opinion that trichocysts are defensive organs; when they come out they form a felt like coating surrounding the organism. It has also been suggested by some that trichocysts act as mooring threads helping in attachment.

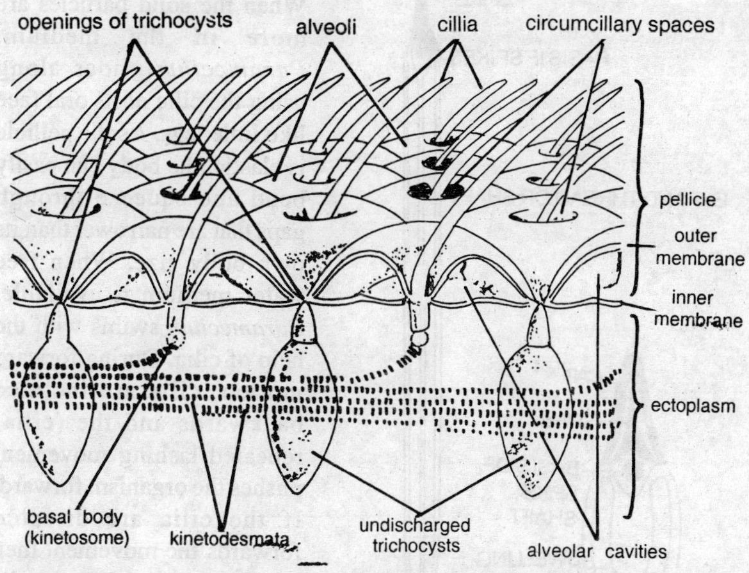

Fig. 12.13 Protozoa
Pellicular system in Paramecium

Endoplasm is granular and less viscous than the ectoplasm. The most prominent component of endoplasm is the nuclear apparatus consisting of a macronucleus (large), and a micronucleus (small). The micronucleus lies in a cleft on the side of the macronucleus. While the macronucleus regulates all the celluar activities, micronucleus seems to be concerned with conjugation (reproduction).

Also present in the endoplasm are a number of food vaculoes. In some strains of *Paramecium,* a number of wedge shaped particles called *kappa particles* are found in the endoplasm. In some cells, these may be more than 400–500 in number. *Kappa particles* are known to produce a poisonous substance called *paramecin,* which when released into the external medium has the capacity to kill other (sensitive) strains of *Paramecium.* Strains of *Paramecia* with kappa particles are called *killer strains,* and those without are called *sensitive strains.*

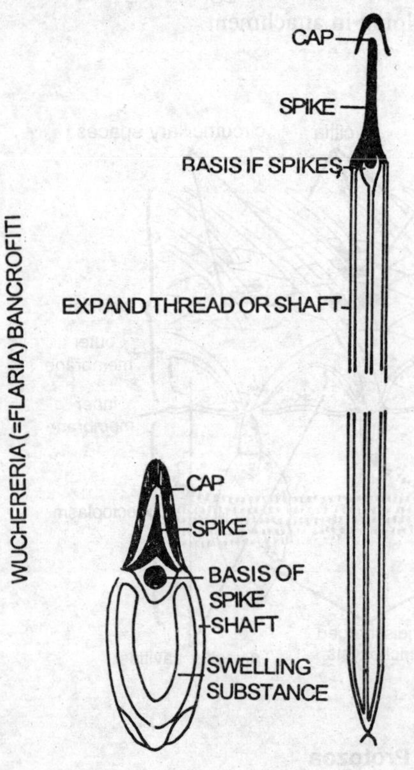

Fig. 12.14 Protozoa
Paramecium. **A.** Undischarged trichocyst, **B.** Discharged trichocyst

Locomotion

Two types of locomotion are seen in *Paramecium*. These are creeping and swimming. When the solid particles are more in the medium, *Paramecium* glides along using the cilia of its oral face like little legs. As the pellicle is elastic, the body can easily bend and squeeze through gaps that are narrower than its own body size. When free water medium is available, *Paramecium* swims with the help of cilia. During forward motion, cilia are inclined backwards and the (cilia) repeated lashing movement pushes the organism forward. If the cilia are directed forwards the movement then will be backward. Whether the movement is forward or backward, it will not be in straight line. Owing to the arrangement of cilia and the twist of the body, the path of movement is in the form of an elongated spiral.

Nutrition

Paramecium exhibits a typical holozoic nutrition. The food consists of mainly smaller protozoa, bacteria and minute particles of plant and animal origin suspended in water. The organism is in constant motion in search of food, and remains in a place where food materials are plenty.

The amount of food required by an individual *Paramecium* varies. When it is growing and actively dividing on an average it requires about 1000–1500 bacteria per day.

Cilia present in the oral groove (peristome) play an important role in capturing the food. The continuous and coordinated lashing of the cilia directs the water along with food particles to move down the gullet. The particles settle down at the bottom of the gullet and eventually form a food vacuole. Soon after the first food vacuole is detached from the gullet, another one begins to form. The detached gastric vacuole is carried round the body due to the streaming movement of cytoplasm called *cyclosis*. During its journey throughout the cell, the food particle gets digested. This whole process may take about three hours. During digestion, the surrounding cytoplasmic medium secretes enzymes into the vacuole. If the food vacuole contains a live prey, an acidic juice is secreted to kill the prey. It is believed that *paramecium* cannot digest fat.

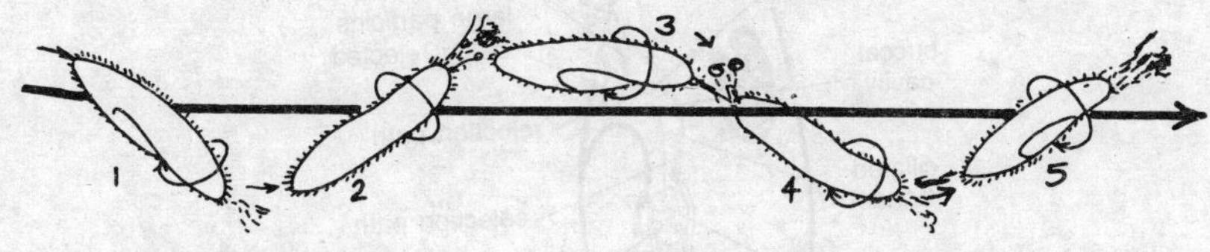

Fig. 12.15 Protozoa
Paramecium Path followed during locomotion

Excretion

Contractile vacuoles which lie at either end of the cell play a prominent role in excretion. The radiating canals that surround the central vesicle of the vacuole, collect the waste matter and surplus water and deliver it to the central vesicle.

The mechanism of contraction in the vacuole is not completely understood. The vacuole perhaps acts like a pump attracting the protoplasm towards it and drives the water out through the surface. It has been noticed that the two vacuoles do not contract simultaneously. They do so alternatively, the interval between the two varying from 10–20 seconds. As in *Amoeba,* in *Paramecium* also, contractile vacuoles perform the function of osmoregulation.

Reproduction

Paramecium reproduces asexually as well as sexually. Asexual reproduction takes place by binary fission, while sexual reproduction is brought about by means of conjugation. A exual reproduction is more frequent than sexual.

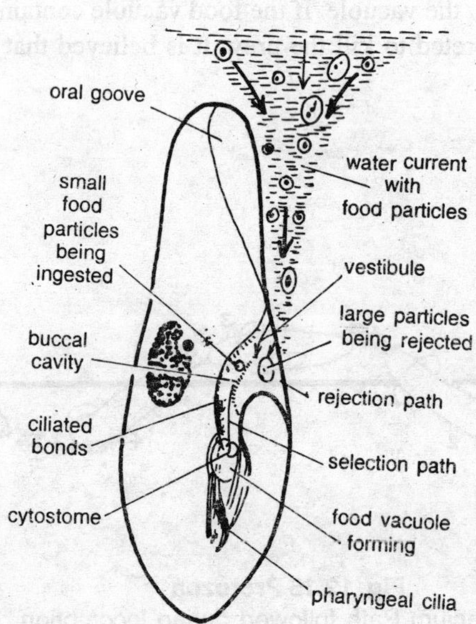

Fig. 12.16 Protozoa
Feeding or ingestion of foodin

Binary fission

This takes place by transverse spliting (transverse binary fission). The first sign of fission is the enlargment of micro nucleus. Chromosomes become visible; they duplicate and migrate to the opposite poles of the micronucleus, which by now becomes greatly elongated. Finally, the micronucleus splits into two and migrates to the opposite ends of the cell. If any piece of micronucleus is left in the centre, it gets absorbed by cytoplasm. At the same time, the macronucleus also divides into two with the two bits migrating to opposite ends. Meanwhile, a constriction appears at the central region of the organism. The division of the macronucleus is amitotic. After the transverse spiliting of the cell, the daughter cell will develop structures which they could not get from the parent cell. Example–after division the oral groove goes to the anterior end daughter cell, while the posterior end daughter cell has to develop a new one on its own.

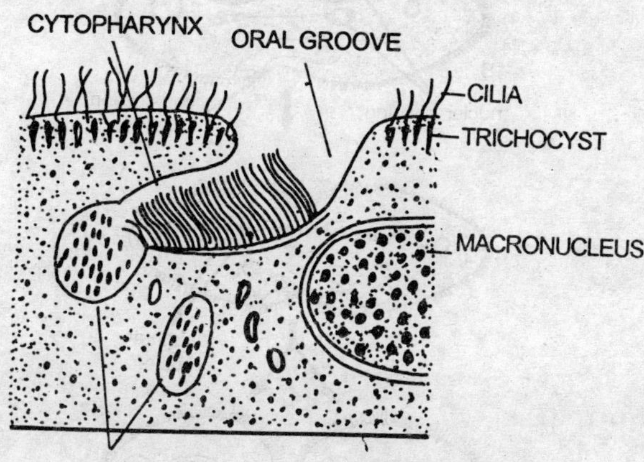

Fig. 12.17 Protozoa
Oral groove with food vacuole under formation

It has been estimated that on an average, a binary fission takes about 2 hours. Temperature and available nutrition play an important role in determining the duration of the division.

Conjugation

Continued binary fission makes the individuals inactive and senility sets in. Reactivation is then brought in by resorting to conjugation which is a means of sexual reproduction.

The first step of conjugation is meeting together of two individuals and their temporary union. There are two mating types or strains in *Paramecia,* and conjugation can take place only between the compatible strains. This has been demonstrated clearly by Sonnerborn T.M., an American scientist.

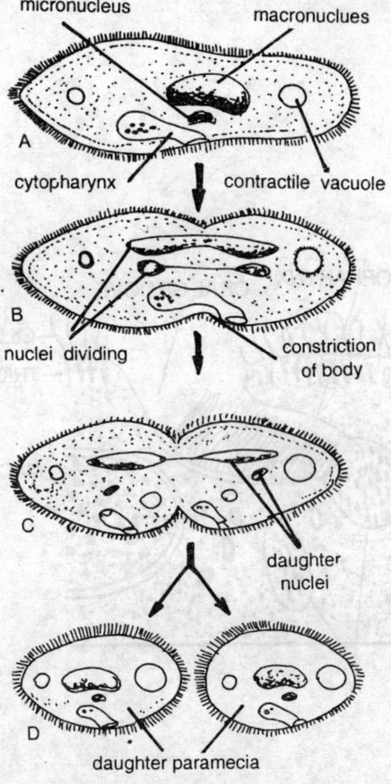

Fig. 12.18 Protozoa
Binary fission in Paramecium

The two individuals that are compatible for mating are called conjugants. The conjugants pair with their ventral side opposite to each other and become united because of the adhesive property of the ectoplasm. Subsequently, they become more intimately associated by their gullets. A little later, the gullets degenerate and a protoplasmic bridge is established between the conjugants. The paired conjugants retain the power of locomotion. The nuclei undergo several changes which is often compared to the process of maturation or gametogenesis. These nuclear changes are :

(i) Meganucleus gradually becomes indistinct and eventually gets absorbed by the cytoplasm. Meanwhile, the micronucleus enlarges in size and undergoes two mitotic divisions to produce four micronuclei in each conjugant.

(ii) Out of four micronuclei (in each conjugant), three degenerate, the remaining one divides again to form two unequal gamete nuclei or pronuclei.

(iii) Out of the two unequal sized nuclei, the large one remains stationary (in the conjugant where it is produced), while the small nucleus migrates to the other conjugant and fuses with the large stationary nucleus. The migrating nuclei are said to be male and the stationary ones are regarded as female.

(iv) As a result of fusion, each conjugant will have a zygote (fused product) nucleus.

Each zygote nucleus (synkaryon) divides thrice to form eight micronuclei, and while these divisions are taking place the conjugants separate. Four of the eight nuclei enlarge and become marconuclei, and three of the other micronuclei disappear. The remaining one divides to form two micronuclei. At the same time, the exconjugant (the organism which has undergone conjugation) divides by binary fission with each daughter cell having one micronucleus and two micronuclei. The two daughter cells of each exconjugant divide again to form four individuals, and meanwhile the micronuclei divide once again (but not the macronucleus), so that each individual will have one macro nucleus and one micronucleus.
After conjugation is complete, a total of eight individuals are formed. It is believed that one of the divisions during formation of gametic nuclei must be reductional in order to maintain the right number of chromosomes.

According to Hadzi, who has studied the process of conjugation in detail, the conjugants are hermaphrodites, and conjugation is essentially a process of reciprocal fertilization.

In addition to conjugation, *Paramecium* is also known to reproduce by endomixis, and autogamy.

Endomixis

This was first reported by Woodruff and Erdmann in *Paramecium aurelia.* In endomixis, there is nuclear reorganisation within an individual, and there is no involvement of another individual.

In Endomixis, the macronucleus does not take part. It disappears while the two micro nuclei divide twice producing eight nuclei. Of these, six disappear. Simultaneously the cell divides, and each daughter cell will have a single nucleus. The single nucleus in each individual (cell) divides twice to produce four nuclei; of these two form macronuclei and the other two micronuclei

The two *Paramecia* divide again along with the division of only micronuclei, resulting in the formation of four individuals each, with one macronucleus and two micronuclei.

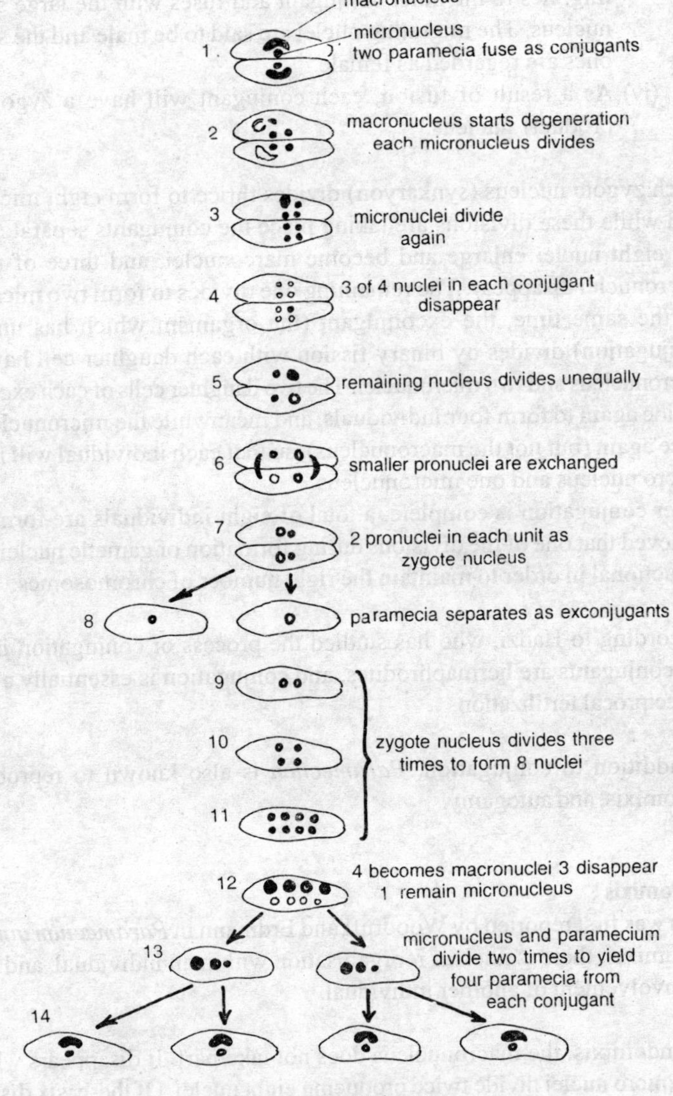

Fig. 12.19 Protozoa
Conjugation in Paramecium caudatum

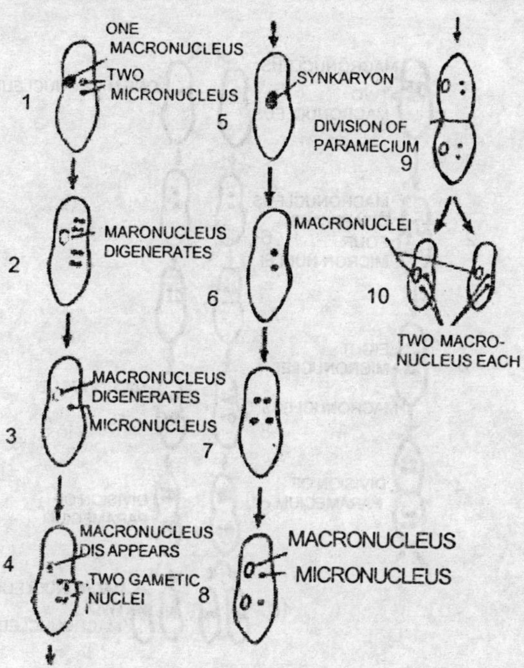

Fig. 12.20 Protozoa
Stages in Endomixis

Autogamy (Automixis)
This is somewhat similar to endomixis, in the sense the entire process occurs in a single individual. Here also, the macronucleus dissappears at the beginning.

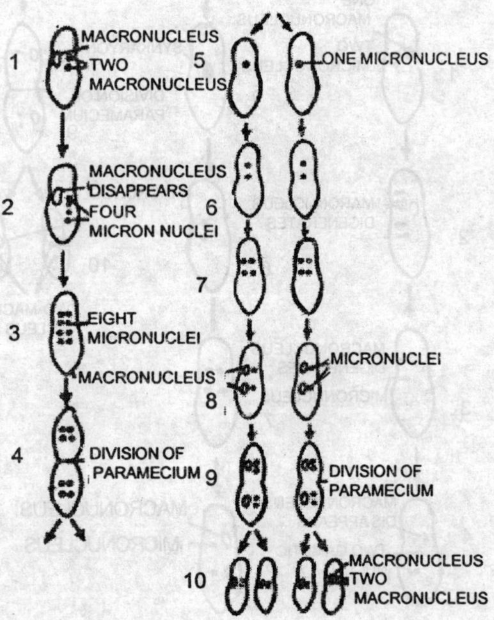

Fig. 12.21 Protozoa
Stages in Autogamy

In autogamy, the two micronuclei divide twice (one of which is meiotic), resulting in the formation of eight nuclei of which seven degenerate.

The remaining one divides once to form two nuclei. These are called the gametic nuclei; they fuse to form a zygote nucleus. This diploid nucleus divides twice to form four nuclei, of which two become micronuclei and the other two macronuclei. The cell (individual) itself divides now followed by one division of micronuclei to form two Paramecia each, with one macronucleus and two micronuclei.